lectures on
quantum mechanics
simple systems

lectures on quantum mechanics

simple systems

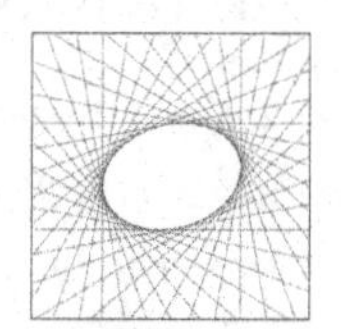

Berthold-Georg Englert
National University of Singapore, Singapore

World Scientific

NEW JERSEY • LONDON • SINGAPORE • BEIJING • SHANGHAI • HONG KONG • TAIPEI • CHENNAI

Published by

World Scientific Publishing Co. Pte. Ltd.

5 Toh Tuck Link, Singapore 596224

USA office: 27 Warren Street, Suite 401-402, Hackensack, NJ 07601

UK office: 57 Shelton Street, Covent Garden, London WC2H 9HE

British Library Cataloguing-in-Publication Data
A catalogue record for this book is available from the British Library.

LECTURES ON QUANTUM MECHANICS
(In 3 Volumes)
Volume 2: Simple Systems

ISBN-13 978-981-256-790-1 (Set)
ISBN-10 981-256-790-9 (Set)
ISBN-13 978-981-256-791-8 (pbk) (Set)
ISBN-10 981-256-791-7 (pbk) (Set)

ISBN-13 978-981-256-972-1 (Vol. 2)
ISBN-10 981-256-972-3 (Vol. 2)
ISBN-13 978-981-256-973-8 (pbk) (Vol. 2)
ISBN-10 981-256-973-1 (pbk) (Vol. 2)

Printed in Singapore

To my teachers, colleagues, and students

Preface

This book on the quantum mechanics of *Simple Systems* grew out of a set of lecture notes for a third-year undergraduate course at the National University of Singapore (NUS). The reader is expected to have the minimal knowledge of a standard brief introduction to quantum mechanics with its typical emphasis on one-dimensional position wave functions.

Proceeding from there, Dirac's formalism of kets, bras, and all that is introduced immediately. In this natural language of the trade, the elementary situations of no force, constant force, and linear restoring force are then dealt with in considerable detail, with Schrödinger's and Heisenberg's equations of motion on equal footing. After treating orbital angular momentum and hydrogen-like atoms, there follows a final chapter on approximation methods, from the Hellmann–Feynman theorem to the WKB quantization rule. For the benefit of the learning student, intermediate steps are not skipped and dozens of exercises are incorporated into the text.

Two companion books on *Basic Matters* and *Perturbed Evolution* cover the material of the preceding and subsequent courses at NUS for second- and fourth-year students, respectively. The three books are, however, not strictly sequential but rather independent of each other and largely self-contained. In fact, there is quite some overlap and a considerable amount of repeated material. While the repetitions send a useful message to the self-studying reader about what is more important and what is less, one could do without them and teach most of *Basic Matters*, *Simple Systems*, and *Perturbed Evolution* in a coherent two-semester course on quantum mechanics.

All three books owe their existence to the outstanding teachers, colleagues, and students from whom I learned so much. I dedicate these lectures to them.

I am grateful for the encouragement of Professors Choo Hiap Oh and Kok Khoo Phua who initiated this project. The professional help by the staff of World Scientific Publishing Co. was crucial for the completion; I acknowledge the invaluable support of Miss Ying Oi Chiew and Miss Lai Fun Kwong with particular gratitude. But nothing would have come about, were it not for the initiative and devotion of Miss Jia Li Goh who turned the original handwritten notes into electronic files that I could then edit.

I wish to thank my dear wife Ola for her continuing understanding and patience by which she is giving me the peace of mind that is the source of all achievements.

Singapore, March 2006 *BG Englert*

Contents

Chapter 1

Quantum Kinematics Reviewed

1.1 Schrödinger's wave function

A typical first course on quantum mechanics is likely to adopt the strategy of the typical textbooks for beginners, and will, therefore, focus predominantly on single objects moving along the x axis. In such an approach, Erwin Schrödinger's wave function $\psi(x)$ plays the central role in the mathematical description of the physical situation. It is taken for granted that the reader is somewhat familiar with the standard material of such a first course.

We remind ourselves of the significance of the wave function $\psi(x)$: by integrating the squared modulus of $\psi(x)$, you get probabilities. In particular, we recall that

$$\int_a^b \mathrm{d}x\, |\psi(x)|^2 \quad \text{is the probability of finding the object between } x = a \text{ and } x = b \text{ (whereby } a < b\text{)}, \tag{1.1.1}$$

which is graphically represented by

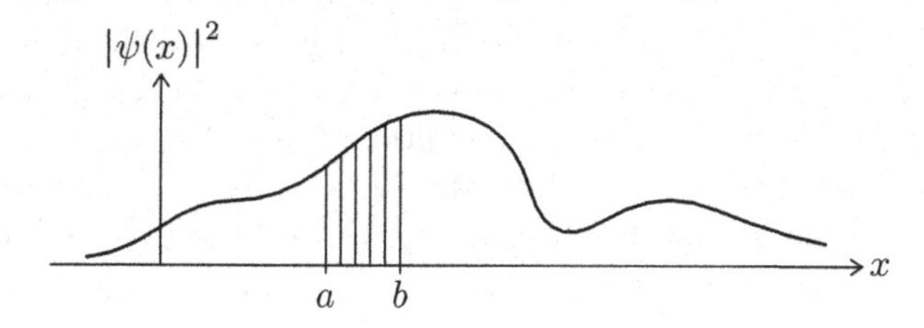

where the area is (proportional to) that probability. Therefore, the squared wave function $|\psi(x)|^2$ is a *probability density*, and one refers to the wave function itself as a *probability density amplitude*. Since we shall surely find the object somewhere, we have unit probability (= 100%) in the limit of $a \to -\infty$, $b \to \infty$, so that $\psi(x)$ is normalized in accordance with

$$\int_{-\infty}^{\infty} \mathrm{d}x \, |\psi(x)|^2 = 1 \,. \tag{1.1.2}$$

We note that these probabilities are of a fundamental nature, they do not result from a lack of knowledge, as it would be typical for the probabilities in classical statistical physics. Also, one must remember that it is the *sole* role of $\psi(x)$ to supply the probabilistic predictions, it has no other significance beyond that. In particular, it would be wrong to think of $|\psi(x)|^2$ as a statement of how the object (electron, atom, ...) is spread out in space. Electrons, and atoms for the present matter as well, are point-like objects. You look for them, and you find them in one place and in one piece. It is only that we cannot predict with certainty the outcome of such a position measurement of an electron. What we can predict reliably are the probabilities of finding the electron in certain regions. And by repeating the measurement very often, we can verify such statistical predictions experimentally.

"Repeating the measurement" means "measure again on an equally prepared electron", it does not mean "measure again the position of the same electron". In the latter situation, the second measurement has probabilities different from the first measurement because the first measurement involves an interaction with the electron and thus a disturbance of the electron. In short, after the first position measurement, there is an altered wave function from which the probabilities for the second measurement are to be derived.

This last remark is a reminder that all probabilities are conditional probabilities. We make statistical predictions based on what we know on the conditions under which the experiment is performed. When speaking of "equally prepared electrons" we mean that the same conditions are realized. After the measurement has been carried out, the conditions are changed, and we must update our statistical predictions accordingly, because the altered conditions determine the probabilities of subsequent measurements.

We recall further that, in addition to the *position wave function* $\psi(x)$, there is also a *momentum wave function* $\psi(p)$, and the two are related to each other by Fourier transformations,

$$\psi(p) = \int \mathrm{d}x \, \frac{\mathrm{e}^{-\mathrm{i}px/\hbar}}{\sqrt{2\pi\hbar}} \psi(x) \,, \qquad \psi(x) = \int \mathrm{d}p \, \frac{\mathrm{e}^{\mathrm{i}xp/\hbar}}{\sqrt{2\pi\hbar}} \psi(p) \,, \tag{1.1.3}$$

where $\hbar$ is Planck's constant (Max K. E. L. Planck). We get probabilities for momentum measurements by integrating $|\psi(p)|^2$,

$$\int_q^r dp\,|\psi(p)|^2 = \text{probability of finding the object's momentum in the range } q < p < r\,, \tag{1.1.4}$$

and

$$\int_{-\infty}^{\infty} dp\,|\psi(p)|^2 = 1 \tag{1.1.5}$$

is the appropriate normalization of $\psi(p)$.

As Jean B. J. Fourier's taught us, the two transformations in (1.1.3) are inverses of each other, so that we can go back and forth between $\psi(x)$ and $\psi(p)$. Their one-to-one correspondence tells us that either one contains all the information of the other. And it does not stop here. For example, we could keep a record of all the gaussian moments of $\psi(x)$,

$$\psi_n = \int dx\,\psi(x)x^n\,e^{-(x/a)^2} \tag{1.1.6}$$

with an arbitrary length parameter a, and the set of $\psi_n, n = 0, 1, 2, \ldots$, would specify $\psi(x)$ uniquely. Clearly, then, the wave functions $\psi(x)$ and $\psi(p)$, and the moments ψ_n are just particular parameterizations of the given *physical state of affairs.* We are thus invited to look for a more abstract entity, a mathematical object that we shall call the *state of the system.*

1.2 Digression: Vectors, coordinates, and all that

It helps to build on an analogy that you might want to keep in mind because it will be useful for the visualization of some rather abstract quantum mechanical statements in terms of geometrical objects. We consider real n-component vectors and their numerical description in terms of coefficients (coordinates) that refer to agreed-upon coordinate systems:

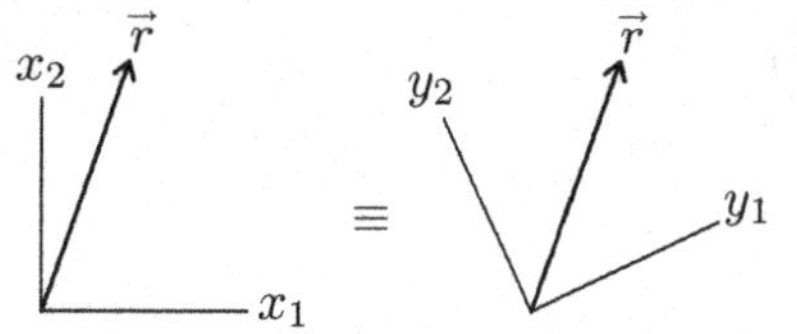

$$\vec{r} \mathrel{\hat{=}} (x_1, x_2, \ldots, x_n) \mathrel{\hat{=}} (y_1, y_2, \ldots, y_n)\,. \tag{1.2.1}$$

One and the same vector has two (or more) numerical descriptions, the coordinates x_j and the coordinates y_k. These numbers, although not unrelated, can be quite different, but they mean *the same vector* $\vec{r}$. We make this explicit with the aid of the basis vectors $\vec{e}_j$ (for the x description) and $\vec{f}_k$ (for the y description),

$$\vec{r} = \sum_{j=1}^{n} x_j \vec{e}_j = \sum_{k=1}^{n} y_k \vec{f}_k \,. \tag{1.2.2}$$

We take for granted (this is a matter of convenient simplicity, not one of necessity) that the basis vectors of each set are orthonormal,

$$\begin{aligned} \vec{e}_j \cdot \vec{e}_k &= \delta_{jk} = \begin{cases} 1 & \text{if} \quad j = k\,, \\ 0 & \text{if} \quad j \neq k\,, \end{cases} \\ \vec{f}_j \cdot \vec{f}_k &= \delta_{jk}\,, \end{aligned} \tag{1.2.3}$$

where δ_{jk} is Leopold Kronecker's delta symbol. Then

$$x_j = \vec{e}_j \cdot \vec{r} \quad \text{and} \quad y_k = \vec{f}_k \cdot \vec{r} \tag{1.2.4}$$

tell us how we determine the coordinates of $\vec{r}$ if the basis vectors are given.

More implicitly than explicitly we have been thinking of $\vec{r}$, $\vec{e}_j$, $\vec{f}_k$ as being numerically represented by *rows* of coordinates. So let us regard the vectors themselves as *row-type vectors.* But just as well we could have arranged the coordinates in columns and would then regard the vectors themselves as *column-type vectors.* It is expedient to emphasize the row or column nature by the notation. We continue to write $\vec{r}$, $\vec{e}_j$, $\vec{f}_k$ for the row vectors, and denote the corresponding column vectors by $r^{\downarrow}$, $e_j^{\downarrow}$, $f_k^{\downarrow}$. Thus

$$\vec{r} \mathrel{\widehat{=}} (x_1, x_2, \ldots, x_n) \mathrel{\widehat{=}} (y_1, y_2, \ldots, y_n) \tag{1.2.5}$$

is paired with

$$r^{\downarrow} \mathrel{\widehat{=}} \begin{pmatrix} x_1 \\ x_2 \\ \vdots \\ x_n \end{pmatrix} \mathrel{\widehat{=}} \begin{pmatrix} y_1 \\ y_2 \\ \vdots \\ y_n \end{pmatrix} \tag{1.2.6}$$

and the two kinds of vectors are related to each other by *transposition,*

$$r^{\downarrow} = \vec{r}^{\,\mathrm{T}}\,, \quad \vec{r} = r^{\downarrow \mathrm{T}}\,. \tag{1.2.7}$$

One immediate benefit of distinguishing between row vectors and column vectors is that we can write inner (scalar, dot) products as simple column-times-row products. This is illustrated by

$$\vec{r} \mathrel{\widehat{=}} (x_1, \ldots, x_n)\,, \qquad \vec{s} \mathrel{\widehat{=}} (u_1, u_2, \ldots, u_n)\,, \tag{1.2.8}$$

$$\vec{r}\cdot\vec{s} = \sum_j x_j u_j = \underbrace{(x_1, x_2, \ldots, x_n)}_{\widehat{=}\,\vec{r}} \underbrace{\begin{pmatrix} u_1 \\ u_2 \\ \vdots \\ u_n \end{pmatrix}}_{\widehat{=}\, s^{\downarrow}} = \vec{r}\, s^{\downarrow}\,. \tag{1.2.9}$$

In view of the symmetry $\vec{r}\cdot\vec{s} = \vec{s}\cdot\vec{r}$, we thus have

$$\underbrace{\vec{r}\cdot\vec{s}}_{\substack{\text{inner product of}\\ \text{two row vectors}}} = \underbrace{\vec{r}\, s^{\downarrow} = \vec{s}\, r^{\downarrow}}_{\substack{\text{products of}\\ \text{the type}\\ \text{``row times}\\ \text{column''}}} = \underbrace{r^{\downarrow}\cdot s^{\downarrow}}_{\substack{\text{inner product of two}\\ \text{column vectors}}}\,. \tag{1.2.10}$$

The central identity here is, of course, consistent with the product rule for transposition, generally: $(AB)^{\mathrm{T}} = B^{\mathrm{T}}A^{\mathrm{T}}$, here:

$$(\vec{r}\, s^{\downarrow})^{\mathrm{T}} = s^{\downarrow\mathrm{T}}\vec{r}^{\,\mathrm{T}} = \vec{s}\, r^{\downarrow}\,, \quad \text{indeed}\,. \tag{1.2.11}$$

Upon combining

$$\vec{r} = \sum_j x_j \vec{e}_j \quad \text{and} \quad x_j = \vec{e}_j\cdot\vec{r} = \vec{r}\, e_j^{\downarrow} \tag{1.2.12}$$

into

$$\vec{r} = \sum_j \vec{r}\, e_j^{\downarrow}\vec{e}_j = \vec{r}\sum_j e_j^{\downarrow}\vec{e}_j \tag{1.2.13}$$

we meet an object of a new kind, the sum of products $e_j^{\downarrow}\vec{e}_j$ of "column times row" type. That is not a number but a *dyadic*, which would have a $n \times n$ matrix as its numerical representation. See, for example,

$$r^{\downarrow}\vec{s} \mathrel{\widehat{=}} \begin{pmatrix} x_1 \\ \vdots \\ x_n \end{pmatrix} (u_1, \ldots, u_n) = \begin{pmatrix} x_1u_1 & x_1u_2 & \cdots & x_1u_n \\ x_2u_1 & x_2u_2 & \cdots & x_2u_n \\ \vdots & \vdots & \ddots & \vdots \\ x_nu_1 & x_nu_2 & \cdots & x_nu_n \end{pmatrix}. \tag{1.2.14}$$

The particular dyadic that appears in (1.2.13) has the property that when it multiplies (on the left) the arbitrary vector $\vec{r}$, the outcome is this vector itself: it is the *unit dyadic*,

$$\vec{r}\,\overleftrightarrow{1} = \vec{r} \quad \text{with} \quad \overleftrightarrow{1} = \sum_j e_j^{\downarrow}\vec{e}_j\,. \tag{1.2.15}$$

The notation $\overleftrightarrow{}$ reminds us that such a dyadic is like a column vector on the left, and like a row vector on the right.

The identification of the unit dyadic is consistent only if it also acts accordingly on the right. Indeed, it does,

$$\begin{aligned} \overleftrightarrow{1}\,r^{\downarrow} &= \sum_j e_j^{\downarrow}\vec{e}_j r^{\downarrow} = \sum_j e_j^{\downarrow}\vec{e}_j\cdot\vec{r} \\ &= \sum_j e_j^{\downarrow}x_j = r^{\downarrow}\,. \end{aligned} \tag{1.2.16}$$

As this little calculation demonstrates, the statement

$$\sum_j e_j^{\downarrow}\vec{e}_j = \overleftrightarrow{1} \tag{1.2.17}$$

expresses the *completeness* of the set of column vectors $e_j^{\downarrow}$, and also that of the set of row vectors $\vec{e}_j$, because we can expand any arbitrary vector $r^{\downarrow}$ as a linear combination of the $e_j^{\downarrow}$.

The statements of *orthonormality*,

$$\vec{e}_j e_k^{\downarrow} = \delta_{jk}\,, \tag{1.2.18}$$

and of completeness in (1.2.17) are two sides of the same coin. And, of course, there is nothing special here about the $\vec{e}_j$ set of vectors, the $\vec{f}_k$ basis vectors are also orthonormal,

$$\vec{f}_j f_k^{\downarrow} = \delta_{jk} \tag{1.2.19}$$

and complete,

$$\sum_k f_k^{\downarrow}\vec{f}_k = \overleftrightarrow{1}\,. \tag{1.2.20}$$

In

$$r^{\downarrow} = \sum_j e_j^{\downarrow}x_j = \sum_k f_k^{\downarrow}y_k \tag{1.2.21}$$

we have two parameterizations of $r^\downarrow$. How does one express one set of coefficients in terms of the other, that is: How does one translate the x description into the y description and vice versa? That is easy! See,

$$\begin{aligned} x_j = \vec{e}_j r^\downarrow &= \vec{e}_j \sum_k f_k^\downarrow y_k \\ &= \sum_k \vec{e}_j f_k^\downarrow y_k = \sum_k (ef)_{jk} y_k \end{aligned} \tag{1.2.22}$$

with

$$(ef)_{jk} = \vec{e}_j f_k^\downarrow = \vec{e}_j \cdot \vec{f}_k \,; \tag{1.2.23}$$

and likewise

$$y_k = \sum_j (fe)_{kj} x_j \quad \text{with} \quad (fe)_{kj} = \vec{f}_k \cdot \vec{e}_j \,. \tag{1.2.24}$$

The two $n \times n$ transformation matrices composed of the matrix elements $(ef)_{jk}$ and $(fe)_{kj}$ are clearly transposes of each other. Furthermore, it follows from

$$\begin{aligned} x_j = \sum_k (ef)_{jk} y_k &= \sum_k (ef)_{jk} \sum_{j'} (fe)_{kj'} x_{j'} \\ &= \sum_{j'} \left(\sum_k (ef)_{jk} (fe)_{kj'} \right) x_{j'} \end{aligned} \tag{1.2.25}$$

that

$$\sum_k (ef)_{jk} (fe)_{kj'} = \delta_{jj'} \tag{1.2.26}$$

must hold. This is to say that the two transformation matrices are inverses of each other — hardly a surprise.

1-1 Use the definitions of $(ef)_{jk}$ and $(fe)_{kj'}$ to verify this relation directly.

1-2 What appears in $y_k = \sum_{k'} \boxed{?}_{kk'} y_{k'}$ as the result of converting y_k into x_j and then back to y_k? Verify here too that $\boxed{?}_{kk'} = \delta_{kk'}$.

Rather than converting one description into the other, we can ask how the two sets of basis vectors are related to each other. Since both sets are orthonormal and complete, the mapping

$$e_j^{\downarrow} \longrightarrow f_j^{\downarrow} = \overleftrightarrow{O}\, e_j^{\downarrow} \tag{1.2.27}$$

is a *rotation* in the n-dimensional space. Geometric intuition tells us that there must be a unique dyadic $\overleftrightarrow{O}$ that accomplishes this rotation. We find it by multiplying with $\vec{e}_j$ from the right

$$f_j^{\downarrow}\vec{e}_j = \overleftrightarrow{O}\, e_j^{\downarrow}\vec{e}_j\,, \tag{1.2.28}$$

followed by summing over j and exploiting the completeness of the $\vec{e}$ vectors,

$$\sum_j f_j^{\downarrow}\vec{e}_j = \sum_j \overleftrightarrow{O}\, e_j^{\downarrow}\vec{e}_j = \overleftrightarrow{O} \underbrace{\sum_j e_j^{\downarrow}\vec{e}_j}_{=\overleftrightarrow{1}}\,, \tag{1.2.29}$$

with the outcome

$$\overleftrightarrow{O} = \sum_j f_j^{\downarrow}\vec{e}_j\,. \tag{1.2.30}$$

As an exercise, we verify that it has the desired property:

$$\overleftrightarrow{O}\, e_j^{\downarrow} = \sum_k f_k^{\downarrow} \underbrace{\vec{e}_k e_j^{\downarrow}}_{=\delta_{kj}} = \sum_k f_k^{\downarrow}\delta_{kj} = f_j^{\downarrow}\,, \tag{1.2.31}$$

indeed. Further, we note that

$$\vec{f}_k \overleftrightarrow{O} = \sum_j \underbrace{\vec{f}_k f_j^{\downarrow}}_{=\delta_{kj}} \vec{e}_j = \sum_j \delta_{kj}\vec{e}_j = \vec{e}_k\,, \tag{1.2.32}$$

so that the same dyadic $\overleftrightarrow{O}$ also transforms the rows $\vec{f}_k$ into the $\vec{e}_k$s. Together with the transposed statements we thus have

$$\begin{aligned} f_j^{\downarrow} &= \overleftrightarrow{O}\, e_j^{\downarrow}\,, & \vec{f}_k \overleftrightarrow{O} &= \vec{e}_k\,, \\ \vec{f}_j &= \vec{e}_j \overleftrightarrow{O}^{\mathrm{T}}\,, & \overleftrightarrow{O}^{\mathrm{T}} f_k^{\downarrow} &= e_k^{\downarrow} \end{aligned} \tag{1.2.33}$$

with

$$\overleftrightarrow{O} = \sum_k f_k^{\downarrow}\vec{e}_k \quad \text{and} \quad \overleftrightarrow{O}^{\mathrm{T}} = \sum_j e_j^{\downarrow}\vec{f}_j\,. \tag{1.2.34}$$

Here, too, we can iterate the transformations, as in

$$
\begin{aligned}
f_j^{\downarrow} &= \overleftrightarrow{O}\, e_j^{\downarrow} = \overleftrightarrow{O}\,\overleftrightarrow{O}^{\mathrm{T}} f_j^{\downarrow} \\
\text{or}\quad \vec{e}_k &= \vec{f}_k \overleftrightarrow{O} = \vec{e}_k \overleftrightarrow{O}^{\mathrm{T}} \overleftrightarrow{O}\,,
\end{aligned}
\tag{1.2.35}
$$

and conclude that

$$
\overleftrightarrow{O}\,\overleftrightarrow{O}^{\mathrm{T}} = \overleftrightarrow{1} = \overleftrightarrow{O}^{\mathrm{T}} \overleftrightarrow{O}\,. \tag{1.2.36}
$$

Dyadics with this property, namely: the transpose is the inverse,

$$
\overleftrightarrow{O}^{\mathrm{T}} = \overleftrightarrow{O}^{-1}\,, \tag{1.2.37}
$$

are called *orthogonal*, in analogy to the corresponding terminology for orthogonal matrices in linear algebra.

1-3 How are the transformation matrices $(ef)_{jk}$ and $(fe)_{kj}$ related to the orthogonal dyadic $\overleftrightarrow{O}$?

Actually, in linear algebra the most basic definition of an orthogonal transformation is that it leaves *all* inner products unchanged. That is, for any pair of vectors $\vec{r}, \vec{s}$ we should have

$$
\left(\vec{r}\,\overleftrightarrow{O}\right)\cdot\left(\vec{s}\,\overleftrightarrow{O}\right) = \vec{r}\cdot\vec{s}\,, \tag{1.2.38}
$$

and for any pair $r^{\downarrow}, s^{\downarrow}$ we should have

$$
\left(\overleftrightarrow{O}\, r^{\downarrow}\right)\cdot\left(\overleftrightarrow{O}\, s^{\downarrow}\right) = r^{\downarrow}\cdot s^{\downarrow}\,. \tag{1.2.39}
$$

Indeed, upon switching over to row-times-column products, we have

$$
\begin{aligned}
\vec{r}\, s^{\downarrow} &= \vec{r}\,\overleftrightarrow{O}\,\overleftrightarrow{O}^{\mathrm{T}} s^{\downarrow} \quad \text{from (1.2.38)} \\
\text{and}\quad \vec{r}\, s^{\downarrow} &= \vec{r}\,\overleftrightarrow{O}^{\mathrm{T}} \overleftrightarrow{O}\, s^{\downarrow} \quad \text{from (1.2.39),}
\end{aligned}
\tag{1.2.40}
$$

and $\overleftrightarrow{O}\,\overleftrightarrow{O}^{\mathrm{T}} = \overleftrightarrow{O}^{\mathrm{T}}\overleftrightarrow{O} = \overleftrightarrow{1}$ are implied again.

1-4 One term in the sum of (1.2.17) is

$$\overset{\downarrow\rightarrow}{P}_j = e_j^{\downarrow}\vec{e}_j\,.$$

Show that $\overset{\downarrow\rightarrow}{P}_j{}^2 = \overset{\downarrow\rightarrow}{P}_j$. What is, therefore, the geometrical significance of $\overset{\downarrow\rightarrow}{P}_j$? Repeat for

$$\overset{\downarrow\rightarrow}{P}_{jk} = \overset{\downarrow\rightarrow}{P}_j + \overset{\downarrow\rightarrow}{P}_k \quad \text{with} \quad j \neq k\,.$$

1.3 Dirac's kets and bras

We now return to the discussion of $\psi(x), \psi(p), \ldots$ as equivalent numerical descriptions of the same abstract entity, the state of affairs of the physical system under consideration. Following Paul A. M. Dirac, we symbolize the state by a so-called *ket*, for which we write $|\ \rangle$ if we mean just any state (as we do presently) and fill the gap with appropriate labels if we mean one of a specific set of states, such as $|1\rangle$, $|2\rangle$, $|3\rangle$, ... or $|\alpha\rangle$, $|\beta\rangle$, ..., whatever the convenient and fitting labels may be. Mathematically speaking, kets are vectors, elements of a complex vector space, which just says that we can add kets to get new ones, and we can multiply them with complex numbers to get other, related kets. More generally, any linear combination of kets is another ket.

It helps to think of kets as analogs of column-type vectors, and then

$$\begin{aligned} |\ \rangle &= \int \mathrm{d}x\,|x\rangle\psi(x) \\ &= \int \mathrm{d}p\,|p\rangle\psi(p) \end{aligned} \tag{1.3.1}$$

are analogs to the two decompositions of $r^{\downarrow}$ in (1.2.21). There are crucial differences, however. Then we were summing over discrete indices, now we are integrating over the continuous variables x and p that label the kets. Then we were dealing with real objects — the coordinates x_j and y_k are real numbers — now we have complex-valued wave functions, $\psi(x)$ and $\psi(p)$. But otherwise the analogy is rather close and well worth remembering.

A wave function $\psi(x)$ that is large only in a small x region describes an object that is very well localized in the sense that we can reliably predict that we shall find it in this small region. In the limit of ever smaller regions — eventually a single x value, a point — we would get $|\ \rangle \to |x\rangle$

in some sense and, therefore, $|x\rangle$ refers to the situation "object is at x, exactly". This, however, is no longer a real physical situation but rather the overidealized situation of that unphysical limit. As a consequence, ket $|x\rangle$ is not actually associated with a physically realizable state, it is a convenient mathematical fiction. The unphysical nature is perhaps most obvious when we recall that the perfect localization of the overidealized limit would require a control on the quantum object with infinite precision — and this is never available in an actual real-life experiment.

By the same token, ket $|p\rangle$ refers to the overidealized situation of infinitely sharp momentum, again a mathematical fiction, not a physical reality. Both $|x\rangle$ and $|p\rangle$ kets are extremely useful mathematical objects, but one must keep in mind that a physical ket $|\ \rangle$ *always* involves a range of x values and a range of p values. This range may be small, then we have a well-controlled position, or a well-controlled momentum, but it is invariably a finite range.

Kets $|\ \rangle$, $|x\rangle$, $|p\rangle$, ... are analogs of column-type vectors. They have their partners in the so-called *bras* $\langle\ |$, $\langle x|$, $\langle p|$, ..., which are analogs of row-type vectors. When dealing with the real vectors $r^{\downarrow}$, $\vec{r}$, we related the two kinds to each other by transposition, $r^{\downarrow} = \vec{r}^{\,\mathrm{T}}$. Now, however, the "coordinates" — that is the wave functions $\psi(x)$, $\psi(p)$ — are complex-valued. Therefore, mathematical consistency requires that we supplement transposition with complex conjugation, and thus have *hermitian conjugation*, or, as the physicists say, we

$$\text{"take the adjoint":}\quad |\ \rangle^{\dagger} = \langle\ |\,, \qquad \langle\ |^{\dagger} = |\ \rangle\,. \tag{1.3.2}$$

The built-in complex conjugation becomes visible as soon as we take the adjoint of a linear combination,

$$\begin{aligned}\big(|1\rangle\alpha + |2\rangle\beta\big)^{\dagger} &= \alpha^{*}\langle 1| + \beta^{*}\langle 2|\,,\\ \big(\gamma\langle 3| + \delta\langle 4|\big)^{\dagger} &= |3\rangle\gamma^{*} + |4\rangle\delta^{*}\,.\end{aligned} \tag{1.3.3}$$

In particular the adjoint statements to the decompositions of $|\ \rangle$ in (1.3.1) are

$$\langle\ | = \int \mathrm{d}x\,\psi(x)^{*}\langle x| = \int \mathrm{d}p\,\psi(p)^{*}\langle p|\,. \tag{1.3.4}$$

In further analogy with the column-type vectors of Section 1.2, the kets are also endowed with an inner product, so that the vector space of kets is an inner-product space or Hilbert space, the name honoring David Hilbert's contributions. The notation $|1\rangle \cdot |2\rangle$ of the inner product as a "dot product" is, however, not used at all. In the mathematical literature, inner products are commonly written as $(\ ,\ )$, so that the inner product of two kets would appear as $(|1\rangle, |2\rangle)$ — except that mathematicians are not fond of the ket and bra notation, Dirac's stroke of genius.

Instead, one follows the suggestion of (1.2.10) and understands the inner products of two kets, or two bras, as the analogs of row-times-column products. So the inner product of the ket

$$|1\rangle = \int \mathrm{d}x\, |x\rangle \psi_1(x) \tag{1.3.5}$$

with the ket

$$|2\rangle = \int \mathrm{d}x\, |x\rangle \psi_2(x) \tag{1.3.6}$$

is obtained by multiplying the bra

$$\langle 1| = \int \mathrm{d}x\, \psi_1(x)^* \langle x| \tag{1.3.7}$$

with the ket $|2\rangle$:

$$\begin{aligned}\langle 1|2\rangle &= \int \mathrm{d}x\, \psi_1(x)^* \langle x| \int \mathrm{d}x'\, |x'\rangle \psi_2(x') \\ &= \int \mathrm{d}x\, \psi_1(x)^* \int \mathrm{d}x'\, \langle x|x'\rangle \psi_2(x'),\end{aligned} \tag{1.3.8}$$

where the integration variable of (1.3.6) is changed to x' to avoid confusion. This is also the inner product of bras $\langle 1|$ and $\langle 2|$. As anticipated in (1.3.8), one writes only one vertical line in the bra-ket product $\langle 1|\,|2\rangle = \langle 1|2\rangle$ of bra $\langle 1|$ and ket $|2\rangle$ and speaks of a Dirac bracket or simply *bracket*.

In accordance with what the reader learned in whichever first course on quantum mechanics, we expect the inner product (1.3.8) to be given by

$$\langle 1|2\rangle = \int \mathrm{d}x\, \psi_1(x)^* \psi_2(x), \tag{1.3.9}$$

so that we need $\langle x|x'\rangle$ such that

$$\int \mathrm{d}x'\, \langle x|x'\rangle \psi_2(x') = \psi_2(x) \tag{1.3.10}$$

for all $\psi_2(x)$. Thus, we infer

$$\langle x|x'\rangle = \delta(x - x') \tag{1.3.11}$$

which is to say that x kets and bras are pairwise orthogonal and normalized to the Dirac δ function. There is a longer discussion of the δ function in Section 4.1 of *Basic Matters*, and so we are content here with recalling the basic, defining property, namely,

$$\int \mathrm{d}x'\, \delta(x-x')f(x') = f(x) \tag{1.3.12}$$

for all the functions that are continuous near x.

The normalization (1.1.2) of the wave function,

$$\int \mathrm{d}x\, |\psi(x)|^2 = 1\,, \tag{1.3.13}$$

now appears as

$$\langle\ |\ \rangle = 1\,, \tag{1.3.14}$$

see:

$$\begin{aligned}\langle\ |\ \rangle &= \int \mathrm{d}x\, \psi(x)^* \langle x| \int \mathrm{d}x'\, |x'\rangle \psi(x') \\ &= \int \mathrm{d}x\, \psi(x)^* \underbrace{\int \mathrm{d}x'\, \underbrace{\langle x|x'\rangle}_{=\delta(x-x')} \psi(x')}_{=\psi(x)} \\ &= \int \mathrm{d}x\, |\psi(x)|^2\,. \end{aligned} \tag{1.3.15}$$

In other words, the physical kets $|\ \rangle$, and the physical bras $\langle\ |$, are of *unit length.*

We recall also the physical significance of the bracket $\langle 1|2\rangle$ in (1.3.9), after which it is named *probability amplitude*: Its squared modulus $|\langle 1|2\rangle|^2$ is the probability prob($2 \to 1$) of finding the system in state 1, described by ket $|1\rangle$ and parameterized by the wave function $\psi_1(x)$, if the system is known to be in state 2, with ket $|2\rangle$ and wave function $\psi_2(x)$. We should not fail to note the symmetry possessed by prob($2 \to 1$),

$$\text{prob}(2 \to 1) = |\langle 1|2\rangle|^2 = \text{prob}(1 \to 2)\,, \tag{1.3.16}$$

which is an immediate consequence of

$$\langle 1|2\rangle = \langle 2|1\rangle^*\,, \tag{1.3.17}$$

demonstrated by interchanging the labels, $1 \leftrightarrow 2$, in (1.3.9). The fundamental symmetry (1.3.16) is quite remarkable because it states that the probability of finding $|1\rangle$ when $|2\rangle$ is the case is always exactly equal to the probability of finding $|2\rangle$ when $|1\rangle$ is the case, although these probabilities can refer to two very different experimental situations.

There is a basic requirement of consistency in this context, namely that $\text{prob}(2 \to 1) \leq 1$. Indeed, this is ensured by the well-known *Cauchy–Bunyakovsky–Schwarz inequality*, named after Augustin-Louis Cauchy, Viktor Y. Bunyakovsky, and K. Hermann A. Schwarz. This inequality is the subject matter of the following exercise.

1-5 For all bras $\langle a|$ and all kets $|b\rangle$, show that

$$|\langle a|b\rangle|^2 \leq \langle a|a\rangle\langle b|b\rangle\,,$$

and state under which condition the equal sign applies. Conclude that $\text{prob}(2 \to 1) \leq 1$ because the physical bra $\langle 1|$ is normalized in accordance with (1.3.14), and so is the physical ket $|2\rangle$.

The orthonormality statement (1.3.11),

$$\langle x|x'\rangle = \delta(x - x')\,, \tag{1.3.18}$$

is the obvious analog of (1.2.18),

$$\vec{e}_j e_k^{\downarrow} = \delta_{jk}\,. \tag{1.3.19}$$

We expect that the analog of the completeness relation (1.2.17),

$$\sum_j e_j^{\downarrow} \vec{e}_j = \overset{\leftrightarrow}{1}\,, \tag{1.3.20}$$

reads

$$\int \mathrm{d}x\, |x\rangle\langle x| = 1\,. \tag{1.3.21}$$

Strictly speaking, the symbol on the right is the identity operator — the operator analog of the unit dyadic $\overset{\leftrightarrow}{1}$ — but we will not be pedantic about it and write it just like the number 1. It will always be unambiguously clear from the context whether we mean the unit operator or the unit number. Likewise, the symbol 5, say, can mean the number 5 or 5 times the unit operator, depending on the context. For instance, in $5\langle x| = \langle x|5$ we have the number 5 on the left and 5 times the unit operator on the right. There will be no confusion arising from this convenience in notation.

But we must not forget to verify the *completeness relation* (1.3.21). "Verification" means here just the check that it is consistent with everything else we have so far. For example, is it true that $1|\ \rangle = |\ \rangle$? We check:

$$\begin{aligned}1|\ \rangle &= \underbrace{\left(\int \mathrm{d}x\,|x\rangle\langle x|\right)}_{=1}\underbrace{\int \mathrm{d}x'\,|x'\rangle\psi(x')}_{=|\ \rangle} \\ &= \int \mathrm{d}x\,|x\rangle \int \mathrm{d}x'\,\underbrace{\langle x|x'\rangle}_{=\delta(x-x')}\psi(x') \\ &= \int \mathrm{d}x\,|x\rangle\psi(x) = |\ \rangle, \quad \text{indeed}. \end{aligned} \tag{1.3.22}$$

Similarly, we check that $\langle\ |1 = \langle\ |$. Another little calculation:

$$\begin{aligned}\langle 1|2\rangle &= \langle 1|1|2\rangle \\ &= \int \mathrm{d}x\,\psi_1(x)^*\langle x| \int \mathrm{d}x'\,|x'\rangle\langle x'| \int \mathrm{d}x''\,|x''\rangle\psi_2(x'') \\ &= \underbrace{\int \mathrm{d}x\,\psi_1(x)^* \int \mathrm{d}x'\,\underbrace{\langle x|x'\rangle}_{\delta(x-x')=}\underbrace{\int \mathrm{d}x''\,\underbrace{\langle x'|x''\rangle}_{=\delta(x'-x'')}\psi_2(x'')}_{=\psi_2(x')}}_{=\psi_2(x)} \\ &= \int \mathrm{d}x\,\psi_1(x)^*\psi_2(x)\,, \quad \text{all right as well.}\end{aligned} \tag{1.3.23}$$

Finally, is $1^2 = 1$? Let us see,

$$\begin{aligned}1^2 &= \int \mathrm{d}x\,|x\rangle\langle x| \int \mathrm{d}x'\,|x'\rangle\langle x'| \\ &= \int \mathrm{d}x\,|x\rangle \int \mathrm{d}x'\,\underbrace{\langle x|x'\rangle}_{=\delta(x-x')}\langle x'| \\ &= \int \mathrm{d}x\,|x\rangle\langle x| = 1\,, \quad \text{indeed.}\end{aligned} \tag{1.3.24}$$

In summary, we have for the position states

$$\begin{aligned}
\text{adjoint relations: } & |x\rangle = \langle x|^\dagger\,, \quad \langle x| = |x\rangle^\dagger\,, \\
\text{orthonormality: } & \langle x|x'\rangle = \delta(x-x')\,, \\
\text{completeness: } & \int \mathrm{d}x\,|x\rangle\langle x| = 1\,.
\end{aligned} \tag{1.3.25}$$

And, by the same token, the corresponding statements hold for the momentum states,

$$\begin{aligned}
\text{adjoint relations: } & |p\rangle = \langle p|^\dagger\,, \quad \langle p| = |p\rangle^\dagger\,, \\
\text{orthonormality: } & \langle p|p'\rangle = \delta(p-p')\,, \\
\text{completeness: } & \int \mathrm{d}p\,|p\rangle\langle p| = 1\,,
\end{aligned} \tag{1.3.26}$$

because we can repeat the whole line of reasoning with labels x consistently replaced by labels p.

1.4 *xp* transformation function

Since the two integrals in (1.3.1) are different parameterization for the same ket $|\;\rangle$, there must be well defined relations between the wave functions $\psi(x)$ and $\psi(p)$ and also between the kets $|x\rangle$ and $|p\rangle$. For the wave functions, the relations are the Fourier transformations of (1.1.3). We use them now to establish the corresponding statements that relate $|x\rangle$ and $|p\rangle$ to each other.

First note what is already implicit in (1.3.22), namely that

$$\begin{aligned}
\langle x|\;\rangle &= \langle x|\int \mathrm{d}x'\,|x'\rangle\psi(x') \\
&= \int \mathrm{d}x'\,\underbrace{\langle x|x'\rangle}_{=\delta(x-x')}\psi(x') \\
&= \psi(x)
\end{aligned} \tag{1.4.1}$$

or

$$\psi(x) = \langle x|\;\rangle \tag{1.4.2}$$

and (infer by analogy or repeat the argument)

$$\psi(p) = \langle p|\ \rangle . \tag{1.4.3}$$

Therefore we have

$$\begin{aligned}\langle x|\ \rangle = \psi(x) &= \int \mathrm{d}p\, \frac{\mathrm{e}^{\mathrm{i}xp/\hbar}}{\sqrt{2\pi\hbar}}\psi(p) \\ &= \int \mathrm{d}p\, \frac{\mathrm{e}^{\mathrm{i}xp/\hbar}}{\sqrt{2\pi\hbar}}\langle p|\ \rangle ,\end{aligned} \tag{1.4.4}$$

and this must be true irrespective of the ket $|\ \rangle$ we are considering, so that

$$\langle x| = \int \mathrm{d}p\, \frac{\mathrm{e}^{\mathrm{i}xp/\hbar}}{\sqrt{2\pi\hbar}}\langle p| \tag{1.4.5}$$

follows. The adjoint statement reads

$$|x\rangle = \int \mathrm{d}p\, |p\rangle \frac{\mathrm{e}^{-\mathrm{i}px/\hbar}}{\sqrt{2\pi\hbar}} . \tag{1.4.6}$$

The inverse relations follow from

$$\begin{aligned}\langle p|\ \rangle = \psi(p) &= \int \mathrm{d}x\, \frac{\mathrm{e}^{-\mathrm{i}px/\hbar}}{\sqrt{2\pi\hbar}}\psi(x) \\ &= \int \mathrm{d}x\, \frac{\mathrm{e}^{-\mathrm{i}px/\hbar}}{\sqrt{2\pi\hbar}}\langle x|\ \rangle ,\end{aligned} \tag{1.4.7}$$

which implies

$$\langle p| = \int \mathrm{d}x\, \frac{\mathrm{e}^{-\mathrm{i}px/\hbar}}{\sqrt{2\pi\hbar}}\langle x| \tag{1.4.8}$$

and

$$|p\rangle = \int \mathrm{d}x\, |x\rangle \frac{\mathrm{e}^{\mathrm{i}xp/\hbar}}{\sqrt{2\pi\hbar}} . \tag{1.4.9}$$

They are, of course, all variants of each other. The most basic statement, so far implicit, is that about $\langle x|p\rangle$, the *xp transformation function*:

$$\langle x|p\rangle = \underbrace{\langle x| \int \mathrm{d}x'\, |x'\rangle}_{= \int \mathrm{d}x'\, \delta(x-x')} \frac{\mathrm{e}^{\mathrm{i}x'p/\hbar}}{\sqrt{2\pi\hbar}} = \frac{\mathrm{e}^{\mathrm{i}xp/\hbar}}{\sqrt{2\pi\hbar}} , \tag{1.4.10}$$

that is

$$\langle x|p\rangle = \frac{e^{ixp/\hbar}}{\sqrt{2\pi\hbar}}, \tag{1.4.11}$$

which is the fundamental phase factor of Fourier transformation. It is worth memorizing this expression, as everything else follows from it, sometimes by using the adjoint relation

$$\langle p|x\rangle = \frac{e^{-ixp/\hbar}}{\sqrt{2\pi\hbar}}. \tag{1.4.12}$$

Again, we have the choice of repeating the argument, or we recognize it as a special case of the general statement (1.3.17).

As an illustration of the fundamental role of $\langle x|p\rangle$, we consider,

$$\begin{aligned}\psi(x) &= \langle x|\ \rangle = \langle x|1|\ \rangle \\ &= \langle x|\left(\int dp\,|p\rangle\langle p|\right)|\ \rangle \\ &= \int dp\,\langle x|p\rangle\langle p|\ \rangle \\ &= \int dp\,\frac{e^{ixp/\hbar}}{\sqrt{2\pi\hbar}}\psi(p),\end{aligned} \tag{1.4.13}$$

which takes us back to (1.1.3). Another application is

$$\begin{aligned}\delta(x-x') &= \langle x|x'\rangle = \langle x|1|x'\rangle \\ &= \langle x|\left(\int dp\,|p\rangle\langle p|\right)|x'\rangle \\ &= \int dp\,\langle x|p\rangle\langle p|x'\rangle \\ &= \int dp\,\frac{e^{ixp/\hbar}}{\sqrt{2\pi\hbar}}\frac{e^{-ipx'/\hbar}}{\sqrt{2\pi\hbar}} \\ &= \int \frac{dp}{2\pi\hbar}\,e^{i(x-x')p/\hbar}\end{aligned} \tag{1.4.14}$$

which is the basic Fourier representation of the Dirac δ function. It appears in many forms, all of which are variants of

$$\int dk\,e^{ikx} = 2\pi\delta(x). \tag{1.4.15}$$

This, too, is an identity that is worth remembering.

The formulation of the position-momentum analog of the orthogonal transformation (1.2.27) requires some care because $|x\rangle \to |p\rangle$ is a mapping between objects of different metrical dimensions. Relations such as (1.3.9) or (1.3.13) tell us that the position wave function $\psi(x) = \langle x|\ \rangle$ has the metrical dimension of the reciprocal square root of a distance ($1/\sqrt{\text{cm}}$, say), and likewise the momentum wave function $\psi(p) = \langle p|\ \rangle$ has the metrical dimension of the reciprocal square root of a momentum ($1/\sqrt{\text{g cm/s}}$, for instance). And since the state ket $|\ \rangle$ is dimensionless, see (1.3.14), the bras $\langle x|$ and $\langle p|$ have these metrical dimensions as well, and so do the kets $|x\rangle$ and $|p\rangle$.

Therefore it is expedient to work with dimensionless quantities, for which purpose we introduce an arbitrary length scale a. Then, the analog of (1.2.27) reads

$$|x\rangle\sqrt{a} \to |p\rangle\sqrt{\hbar/a} = U|x\rangle\sqrt{a} \quad \text{for} \quad \frac{x}{a} = \frac{p}{\hbar/a}\,, \tag{1.4.16}$$

where we note that $\hbar/a$ is the corresponding momentum scale because Planck's constant has the metrical dimension of length $\times$ momentum. The operator U thus defined is given by

$$\begin{aligned} U = U1 = U\int \mathrm{d}x\,|x\rangle\langle x| &= \int \mathrm{d}x\,U|x\rangle\langle x| \\ &= \int \mathrm{d}x\,|p = x\hbar/a^2\rangle\sqrt{\hbar/a^2}\langle x| \end{aligned} \tag{1.4.17}$$

or, after substituting $x = ta$,

$$U = \int \mathrm{d}t\,|p = t\hbar/a\rangle\sqrt{\hbar}\langle x = ta|\,. \tag{1.4.18}$$

1-6 Verify that U brings about the transformation in (1.4.16). Then evaluate $\langle p|U$. Do you get what you expect?

The hermitian conjugation of (1.3.2) has the same product rule as transposition, see (1.2.11),

$$(AB)^\dagger = B^\dagger A^\dagger\,, \tag{1.4.19}$$

because the complex conjugation that distinguishes the two operations has no effect on the order of multiplication. In particular we have $\big(|p\rangle\langle x|\big)^\dagger =$

$|x\rangle\langle p|$, so that

$$U^\dagger = \int \mathrm{d}t\,|x = ta\rangle\sqrt{\hbar}\langle p = t\hbar/a| \tag{1.4.20}$$

is the adjoint of U. In analogy with the orthogonality statement (1.2.36), we expect

$$UU^\dagger = 1 = U^\dagger U \tag{1.4.21}$$

to hold. Let us verify the left identity:

$$\begin{aligned}
UU^\dagger &= \int \mathrm{d}t\,|p = t\hbar/a\rangle\sqrt{\hbar}\,\underbrace{\langle x = ta|\int \mathrm{d}t'\,|x = t'a\rangle}_{=\int \mathrm{d}t'\,\delta(ta-t'a)}\,\sqrt{\hbar}\langle p = t'\hbar/a| \\
&= \int \mathrm{d}t\,|p = t\hbar/a\rangle\hbar\underbrace{\int \mathrm{d}t'\,\delta(ta-t'a)\langle p = t'\hbar/a|}_{=(1/a)\langle p = t\hbar/a|} \\
&= \int \mathrm{d}p\,|p\rangle\langle p| = 1\,.
\end{aligned} \tag{1.4.22}$$

It is all right indeed, and the right identity in (1.4.21) is demonstrated the same way. At an intermediate step the identity

$$\delta(ta - t'a) = \frac{1}{a}\delta(t - t') \qquad (a > 0) \tag{1.4.23}$$

is used, which is a special case of (5.1.109) in *Basic Matters*.

Operators with the property (1.4.21), that is: their adjoint is their inverse, are called *unitary operators*. They transform sets of kets or bras into equivalent sets, much like a rotation turns sets of vectors of column or row type into equivalent ones, and play a very important role in quantum mechanics.

1.5 Position operator, momentum operator, functions of them

We look for the object (electron, atom, ...) and find it at position x, with the probability of finding it inside a small vicinity around x given by $\mathrm{d}x\,|\psi(x)|^2$. This is then the probability distribution associated with the (random) variable x. Accordingly, the mean value of x is calculated as

$$\overline{x} = \int \mathrm{d}x\,|\psi(x)|^2 x\,, \tag{1.5.1}$$

and the mean value of x^2 as

$$\overline{x^2} = \int \mathrm{d}x \, |\psi(x)|^2 x^2 \,. \tag{1.5.2}$$

More generally we have

$$\overline{x^n} = \int \mathrm{d}x \, |\psi(x)|^2 x^n \tag{1.5.3}$$

for an arbitrary power, and

$$\overline{f(x)} = \int \mathrm{d}x \, |\psi(x)|^2 f(x) \tag{1.5.4}$$

for the mean value of an arbitrary function of the position variable x.

Recalling that $|\psi(x)|^2 = \psi(x)^* \psi(x) = \langle \; |x\rangle\langle x| \; \rangle$, we rewrite these expressions as

$$\begin{aligned}
\overline{x} &= \langle \; \Big| \Big(\int \mathrm{d}x \, |x\rangle x \langle x| \Big) \Big| \; \rangle \,, \\
\overline{x^2} &= \langle \; \Big| \Big(\int \mathrm{d}x \, |x\rangle x^2 \langle x| \Big) \Big| \; \rangle \,, \\
&\;\;\vdots \\
\overline{f(x)} &= \langle \; \Big| \Big(\int \mathrm{d}x \, |x\rangle f(x) \langle x| \Big) \Big| \; \rangle \,,
\end{aligned} \tag{1.5.5}$$

thereby isolating the specific state of the system — bra on the left, ket on the right — from the quantity that we are taking the average of,

$$\begin{aligned}
\overline{x} &\longrightarrow \int \mathrm{d}x \, |x\rangle x \langle x| \equiv X \,, \\
\overline{x^2} &\longrightarrow \int \mathrm{d}x \, |x\rangle x^2 \langle x| \equiv X^2 \,, \\
&\;\;\vdots \\
\overline{f(x)} &\longrightarrow \int \mathrm{d}x \, |x\rangle f(x) \langle x| \equiv f(X) \,.
\end{aligned} \tag{1.5.6}$$

The first line introduces the *position operator* X as the integral of $|x\rangle x \langle x|$, the second line introduces X^2, the square of X, as we can verify,

$$
\begin{aligned}
XX &= \int \mathrm{d}x\, |x\rangle x \langle x| \int \mathrm{d}x'\, |x'\rangle x' \langle x'| \\
&= \int \mathrm{d}x\, |x\rangle x \underbrace{\int \mathrm{d}x'\, \underbrace{\langle x|x'\rangle}_{=\delta(x-x')} x' \langle x'|}_{=x\langle x|} \\
&= \int \mathrm{d}x\, |x\rangle x^2 \langle x| = X^2, \quad \text{indeed}\,.
\end{aligned} \tag{1.5.7}
$$

And so forth, we have

$$X^n = \int \mathrm{d}x\, |x\rangle x^n \langle x| \tag{1.5.8}$$

for the powers of X. Then by linear combinations,

$$f(X) = \int \mathrm{d}x\, |x\rangle f(x) \langle x| \tag{1.5.9}$$

for all polynomial functions of x and, by approximation, finally for all reasonable functions of x. "Reasonable" means here that the numbers $f(x)$ have to be well defined function values for all real numbers x. Once we have gone through this argument we can just accept (1.5.9) as the definition of a function of position operator X. The integral on the right-hand side of (1.5.9) is an example for the *spectral decomposition* of an operator, here of $f(X)$.

1-7 Show that the eigenvalue equation

$$f(X)|x\rangle = |x\rangle f(x) \tag{1.5.10}$$

holds. What is $\langle x|f(X)$?

1-8 Consider $f(X)^\dagger = \int \mathrm{d}x\, |x\rangle f(x)^* \langle x|$ and compare $f(X)f(X)^\dagger$ with $f(X)^\dagger f(X)$. Which property must be possessed by $f(x)$ if $f(X)$ is its own adjoint, $f(X)^\dagger = f(X)$? Operators with this property are called *selfadjoint* or simply *hermitian* (Charles Hermite), whereby we ignore the subtle difference between the two terms in the mathematical literature. Conclude that all expectation values of a hermitian operator are real. This reality property can serve as an alternative definition of what is a hermitian operator.

Likewise, we have the *momentum operator* P,

$$P = \int \mathrm{d}p\,|p\rangle p\langle p|\,, \tag{1.5.11}$$

and can rely on the spectral decomposition

$$g(P) = \int \mathrm{d}p\,|p\rangle g(p)\langle p| \tag{1.5.12}$$

for all functions of P that derive from reasonable numerical functions $g(p)$. Once again, the reasoning is completely analogous and we need not repeat it.

1-9 Show the following fundamental relations:

$$\langle x|P = \frac{\hbar}{\mathrm{i}}\frac{\partial}{\partial x}\langle x|\,, \qquad X|p\rangle = -\frac{\hbar}{\mathrm{i}}\frac{\partial}{\partial p}|p\rangle\,, \tag{1.5.13}$$

and their integral versions

$$\langle x|\,\mathrm{e}^{\mathrm{i}x'P/\hbar} = \langle x+x'|\,, \qquad \mathrm{e}^{\mathrm{i}p'X/\hbar}|p\rangle = |p+p'\rangle\,. \tag{1.5.14}$$

The latter involve the basic unitary operators associated with P and X; more about them in Section 1.8

Now, with the operator functions of X and P at hand, we return to (1.5.5) and note that

$$\overline{x} = \underbrace{\langle\ |X|\ \rangle}\equiv\underbrace{\langle X\rangle} \tag{1.5.15}$$

average of the numbers x

"expectation value" of position operator X

position operator X sandwiched between state bra $\langle\ |$ and state ket $|\ \rangle$

which introduces a new notation, $\langle X\rangle$, that emphasizes the role played by the position operator X. One speaks of the "expectation value", a historical terminology that is, as so often, not fully logically but completely standard.

Similarly, we write $\langle f(X)\rangle$ for the expectation value of the operator-function $f(X)$ and $\langle P\rangle$, $\langle P^2\rangle$, ..., $\langle g(P)\rangle$, for the expectation values of P, P^2, ..., $g(P)$. We have introduced the latter, by analogy, in terms of integrals involving the momentum wave function $\psi(p)$, but we can, of course, also refer to the position wave function $\psi(x)$.

1-10 In particular, establish expressions (you know them from your first course on quantum mechanics) for the expectation values $\langle P\rangle$ and $\langle P^2\rangle$ in terms of $\psi(x)$.

1-11 Determine the normalization constant A for the position wave function

$$\psi(x)=\begin{cases}A\sin(2\pi x/L) & \text{for}\quad -L<x<L\,,\\ 0 & \text{for}\quad |x|>L\,,\end{cases}$$

and then calculate $\langle X\rangle$, $\langle X^2\rangle$, $\langle P\rangle$, and $\langle P^2\rangle$.

1.6 Traces and statistical operators

Given a ket $|1\rangle$ and a bra $\langle 2|$, we can multiply them in either order, thereby getting

$$\begin{aligned}&\text{the bracket } \langle 2|1\rangle\,,\ \text{a number}\,,\\ \text{or}\ &\text{the “ket-bra” } |1\rangle\langle 2|\,,\ \text{an operator}\,.\end{aligned}$$

In the bracket $\langle 2|1\rangle$, the ingredients are no longer identifiable because very many pairs of a bra and a ket have the same number for the bracket. By contrast, given the ket-bra $|1\rangle\langle 2|$, one can identify the ingredient almost uniquely. Therefore, we cannot expect that there could be a mapping from the bracket to the ket-bra, but there is a mapping from the ket-bra to the bracket,

$$|1\rangle\langle 2|\to\langle 2|1\rangle\,.$$

It is called “taking the trace” and we write $\operatorname{tr}\{\cdots\}$ for it,

$$\operatorname{tr}\{|1\rangle\langle 2|\}=\langle 2|1\rangle\,; \tag{1.6.1}$$

read: the trace of $|1\rangle\langle 2|$ is $\langle 2|1\rangle$.

Before proceeding, let us look at the analog for column and row vectors:

$$r^{\downarrow}\vec{s}\to\vec{s}r^{\downarrow}=\vec{s}\cdot\vec{r}\,, \tag{1.6.2}$$

or

$$\begin{pmatrix} x_1 \\ x_2 \\ \vdots \\ x_n \end{pmatrix} (u_1, u_2, \ldots, u_n) = \begin{pmatrix} x_1u_1 & x_1u_2 & \cdots & x_1u_n \\ x_2u_1 & x_2u_2 & \cdots & x_2u_n \\ \vdots & \vdots & \ddots & \vdots \\ x_nu_1 & x_nu_2 & \cdots & x_nu_n \end{pmatrix}$$
$$\to x_1u_1 + x_2u_2 + \cdots + x_nu_n \,, \tag{1.6.3}$$

the diagonal sum of the matrix for $r^{\downarrow}\vec{s}$. Clearly, if you only know the value of this sum, you cannot reconstruct the whole matrix, but given the matrix, you easily figure out the diagonal sum.

The linear structure for kets and bras is inherited by the trace. For example, consider

$$|1\rangle = |1_a\rangle\alpha + |1_b\rangle\beta \,, \tag{1.6.4}$$

and compare the two ways of evaluating $\mathrm{tr}\{|1\rangle\langle 2|\}$,

$$\begin{aligned} \mathrm{tr}\{|1\rangle\langle 2|\} &= \langle 2|1\rangle = \langle 2|1_a\rangle\alpha + \langle 2|1_b\rangle\beta \\ &= \alpha\,\mathrm{tr}\{|1_a\rangle\langle 2|\} + \beta\,\mathrm{tr}\{|1_b\rangle\langle 2|\} \,, \\ \mathrm{tr}\{|1\rangle\langle 2|\} &= \mathrm{tr}\Big\{\big(|1_a\rangle\alpha + |1_b\rangle\beta\big)\langle 2|\Big\} \\ &= \mathrm{tr}\{|1_a\rangle\alpha\langle 2| + |1_b\rangle\beta\langle 2|\} \,. \end{aligned} \tag{1.6.5}$$

This generalizes to

$$\mathrm{tr}\{\alpha A + \beta B\} = \alpha\,\mathrm{tr}\{A\} + \beta\,\mathrm{tr}\{B\} \tag{1.6.6}$$

immediately, wherein A, B are operators and α, β are numbers.

We apply this to expectation values, such as

$$\langle A\rangle = \langle\ |A|\ \rangle \tag{1.6.7}$$

where A is any operator, perhaps a function $f(X)$ of position operator X, or a function $g(P)$ of momentum operator P, or possibly something more complicated, like the symmetrized product $XP + PX$, say. Whatever the nature of operator A, we can read the above statement as

$$\langle A\rangle = \underbrace{\langle\ |A}_{\text{bra}}\ \underbrace{|\ \rangle}_{\text{ket}} = \mathrm{tr}\Big\{\underbrace{|\ \rangle}_{\text{ket}}\ \underbrace{\langle\ |A}_{\text{bra}}\Big\} \tag{1.6.8}$$

or as

$$\langle A \rangle = \underbrace{\langle \; |}_{\text{bra}} \underbrace{A| \; \rangle}_{\text{ket}} = \operatorname{tr}\left\{ \underbrace{A| \; \rangle}_{\text{ket}} \underbrace{\langle \; |}_{\text{bra}} \right\}, \tag{1.6.9}$$

which introduces $| \; \rangle\langle \; |$ as the mathematical object that refers solely to the state of affairs of the physical system,

$$\langle A \rangle = \operatorname{tr}\left\{ \underbrace{| \; \rangle\langle \; |}_{\text{physical state}} \underbrace{A}_{\text{operator considered}} \right\} = \operatorname{tr}\left\{ \underbrace{A|}_{\text{operator considered}} \underbrace{\; \rangle\langle \; |}_{\text{physical state}} \right\}. \tag{1.6.10}$$

What has been achieved here, is the complete separation of the mathematical entities that refer to the physical property (position, momentum, functions of them, such as energy), and to the state of affairs. There is one appropriate operator A for the physical property, irrespective of the particular state that is actually the case, and all that characterizes the actual situation is contained in the ket-bra

$$\rho \equiv | \; \rangle\langle \; | . \tag{1.6.11}$$

This is also an operator, but not one that describes a physical property, rather it is the *statistical operator* that summarizes all statistical aspects of the system as actually prepared. It is common to refer to the statistical operator as the *state operator* or simply the *state* of the physical system.

In particular, we extract probabilities from ρ as illustrated by the probability of finding the object between $x = a$ and $x = b$ $(a < b)$. We recall that

$$\begin{aligned} \operatorname{prob}(a < x < b) &= \int_a^b \mathrm{d}x \, |\psi(x)|^2 \\ &= \int_{-\infty}^{\infty} \mathrm{d}x \, |\psi(x)|^2 \chi_{a,b}(x) , \end{aligned} \tag{1.6.12}$$

where

$$\chi_{a,b}(x) = \begin{cases} 1 & \text{if} \quad a < x < b , \\ 0 & \text{elsewhere,} \end{cases} \tag{1.6.13}$$

is the *mesa function* for the interval $a < x < b$. Thus,

$$\begin{aligned}\mathrm{prob}(a < x < b) &= \langle\ |\Bigl(\int \mathrm{d}x\, |x\rangle\chi_{a,b}(x)\langle x|\Bigr)|\ \rangle \\ &= \langle\ |\chi_{a,b}(X)|\ \rangle \\ &= \mathrm{tr}\{\chi_{a,b}(X)|\ \rangle\langle\ |\} \\ &= \mathrm{tr}\{\chi_{a,b}(X)\rho\}\ , \end{aligned} \tag{1.6.14}$$

where we recognize the expectation value of a function of position operator X, namely $\chi_{a,b}(X)$. Indeed, such a probability is an expectation value, and the argument is easily extended to other probabilities as well.

1-12 Find the operator A whose expectation value is the probability of finding the object in the range $x > 0$. Express A in terms of the *sign function*

$$\mathrm{sgn}(x) = \begin{cases} +1 & \text{for} \quad x > 0\,, \\ -1 & \text{for} \quad x < 0\,. \end{cases}$$

Repeat for the range $x < 0$, and then check that the two operators add up to the identity.

In (1.6.10) we found

$$\langle A\rangle = \mathrm{tr}\{A\rho\} = \mathrm{tr}\{\rho A\}\ , \quad \rho = |\ \rangle\langle\ |\,, \tag{1.6.15}$$

as if the order of multiplication did not matter. In fact, it does not. We demonstrate this by evaluating $\mathrm{tr}\{AB\}$ and $\mathrm{tr}\{BA\}$ for $A = |1\rangle\langle 2|$ and $B = |3\rangle\langle 4|$, which is all we need, because all operators are linear combinations of such ingredients, and we know already that the trace respects this linearity. Therefore, it suffices to consider these special operators. See, then,

$$\begin{aligned}\mathrm{tr}\{AB\} &= \mathrm{tr}\Bigl\{\underbrace{|1\rangle\langle 2|3\rangle}_{\text{ket}}\underbrace{\langle 4|}_{\text{bra}}\Bigr\} = \langle 4|1\rangle\langle 2|3\rangle\,, \\ \mathrm{tr}\{BA\} &= \mathrm{tr}\Bigl\{\underbrace{|3\rangle\langle 4|1\rangle}_{\text{ket}}\underbrace{\langle 2|}_{\text{bra}}\Bigr\} = \langle 2|3\rangle\langle 4|1\rangle\,. \end{aligned} \tag{1.6.16}$$

Indeed, they are the same because the numbers $\langle 4|1\rangle$ and $\langle 2|3\rangle$ can be multiplied in either order without changing the value of the product. Please note that $AB \neq BA$, as a rule, but their traces are the same.

Since the operators A, B themselves can be products, we have the more general rule

$$\mathrm{tr}\{ABC\} = \mathrm{tr}\{CAB\} = \mathrm{tr}\{BCA\} \tag{1.6.17}$$

for products of three factors and analogous statements about four, five, ... factors. All of them are summarized in the observation that

> the value of a trace does not change when the factors in a product are permuted cyclically.

This *cyclic property of the trace* is exploited very often.

So far, letter ρ was just an abbreviation for the ket-bra $|\;\rangle\langle\;|$, the product of the ket and the bra describing the actual state of affairs. Let us now move on and consider a more general situation:

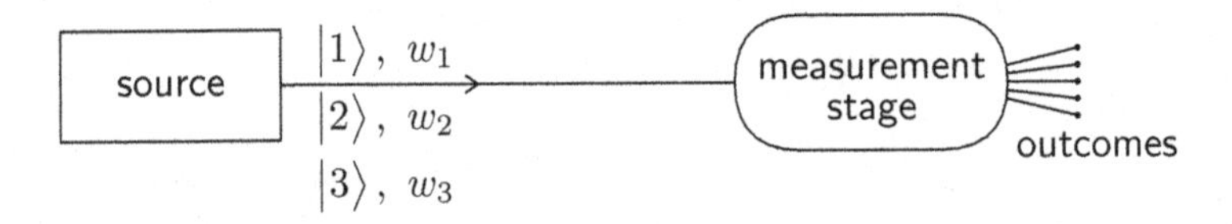

We have a source that puts out the atoms either in state $|1\rangle$, or in state $|2\rangle$, or in state $|3\rangle$, whereby we have no clue what will be the case for the next atom except that we know that there are definite probabilities w_1, w_2, w_3 of occurrence for the three states, with $w_1 + w_2 + w_3 = 1$, of course, because nothing else can possibly happen.

At the measurement stage we perform a measurement of the physical property that is associated with operator A. Thus we have an expectation value that is given by

$$\langle A\rangle = w_1\langle 1|A|1\rangle + \underbrace{w_2}_{\substack{\text{probability}\\ \text{of getting } |2\rangle}} \underbrace{\langle 2|A|2\rangle}_{\substack{\text{expectation value}\\ \text{if } |2\rangle \text{ is the case}}} + w_3\langle 3|A|3\rangle \tag{1.6.18}$$

or

$$\begin{aligned}\langle A\rangle &= w_1\mathrm{tr}\{A|1\rangle\langle 1|\} + w_2\mathrm{tr}\{A|2\rangle\langle 2|\} + w_3\mathrm{tr}\{A|3\rangle\langle 3|\}\\ &= \mathrm{tr}\Big\{A\Big(|1\rangle w_1\langle 1| + |2\rangle w_2\langle 2| + |3\rangle w_3\langle 3|\Big)\Big\}\\ &= \mathrm{tr}\{A\rho\}\end{aligned} \tag{1.6.19}$$

with the statistical operator

$$\rho = \sum_{k=1}^{3} |k\rangle w_k \langle k| \,. \tag{1.6.20}$$

This ρ summarizes all that we know about the source: there are the states $|k\rangle$ and their statistical weights w_k.

There is nothing particular about the situation discussed, with three different states emitted by the source, there could be fewer or more. Accordingly, we have the more general case of

$$\rho = \sum_k |k\rangle w_k \langle k| \,, \quad w_k > 0 \,, \quad \sum_k w_k = 1 \,, \tag{1.6.21}$$

where the summation can have one or more terms, and it is understood that all kets and bras are normalized properly,

$$\langle k|k\rangle = 1 \quad \text{for all } k \,. \tag{1.6.22}$$

By measuring sufficiently many different physical properties, we can establish a body of experimental data that enables us to infer the statistical operator ρ with the desired precision. Then we know the statistical properties of the atoms emitted by the source, and this is *all* we can find out. We cannot, in particular, establish the ingredients $|k\rangle$ and their weights w_k, we can only know ρ. This is also the only really meaningful thing to know since it gives us all probabilities for the statistical predictions. Nothing else is needed. Nor is anything else available.

When we state that knowledge of ρ does not translate into knowledge of the ingredients from which it is composed in

$$\rho = \sum_k |k\rangle w_k \langle k| \,, \tag{1.6.23}$$

we mean of course that different right-hand sides can give the same ρ. To make this point it is quite enough to give one example. The simplest is

$$\rho = \frac{1}{2}\Bigl(|1\rangle\langle 1| + |2\rangle\langle 2|\Bigr) \quad \text{with} \quad \langle 1|2\rangle = 0 \,, \tag{1.6.24}$$

that is just two states mixed with equal weights of 50% for each. In view of their stated orthogonality, the kets

$$|\alpha\rangle = \frac{1}{\sqrt{2}}\Bigl(|1\rangle + |2\rangle\Bigr) \,, \quad |\beta\rangle = \frac{1}{\sqrt{2}}\Bigl(|1\rangle - |2\rangle\Bigr) \tag{1.6.25}$$

are also orthogonal and properly normalized. Then

$$\begin{aligned}&\frac{1}{2}\Bigl(|\alpha\rangle\langle\alpha| + |\beta\rangle\langle\beta|\Bigr)\\ &= \frac{1}{2}\left(\frac{|1\rangle + |2\rangle}{\sqrt{2}}\frac{\langle 1| + \langle 2|}{\sqrt{2}} + \frac{|1\rangle - |2\rangle}{\sqrt{2}}\frac{\langle 1| - \langle 2|}{\sqrt{2}}\right)\\ &= \frac{1}{2}\Bigl(|1\rangle\langle 1| + |2\rangle\langle 2|\Bigr) = \rho \end{aligned} \tag{1.6.26}$$

establishes

$$\rho = \frac{1}{2}\Bigl(|\alpha\rangle\langle\alpha| + |\beta\rangle\langle\beta|\Bigr) \tag{1.6.27}$$

which has different ingredients than the original ρ of (1.6.24). We speak of the two *blends* for one and the same *mixture* or *mixed state*. All that is relevant is the mixture ρ, not the particular ways in which one can blend it.

For

$$\rho = \frac{1}{2}\Bigl(|1\rangle\langle 2| + |2\rangle\langle 2|\Bigr) = \frac{1}{2}\Bigl(|\alpha\rangle\langle\alpha| + |\beta\rangle\langle\beta|\Bigr) \tag{1.6.28}$$

one can say that "it is *as if* we had 50% of $|1\rangle$ and 50% of $|2\rangle$" or one can say with equal justification that "it is *as if* we had 50% of $|\alpha\rangle$ and 50% of $|\beta\rangle$". But neither *as-if reality* is better than the other, both are on exactly the same footing, and there are many more as-if realities associated with this ρ.

1-13 Consider $|u\rangle = \frac{1}{\sqrt{2}}\Bigl(|1\rangle + \mathrm{i}|2\rangle\Bigr)$, $|v\rangle = \frac{1}{\sqrt{2}}\Bigl(|1\rangle - \mathrm{i}|2\rangle\Bigr)$ and evaluate $\frac{1}{2}\Bigl(|u\rangle\langle u| + |v\rangle\langle v|\Bigr)$. What do you conclude?

The basic probability is that of finding a particular state, $|0\rangle$, say. If state $|k\rangle$ is the case, this probability is $\bigl|\langle 0|k\rangle\bigr|^2$, as in (1.3.16), so more generally the probability is $\langle 0|\rho|0\rangle$. It must be nonnegative,

$$\langle 0|\rho|0\rangle \geq 0 \quad \text{for any choice of } |0\rangle\,. \tag{1.6.29}$$

In short: $\rho \geq 0$, which is a basic property of all statistical operators, their *positivity*. Other properties are that ρ is hermitian,

$$\rho = \rho^\dagger\,, \tag{1.6.30}$$

and normalized to unit trace,

$$\mathrm{tr}\{\rho\} = 1\,. \tag{1.6.31}$$

All these properties follow directly from the construction of ρ as a blend in (1.6.21).

We emphasize the physical significance of $\sum_k w_k = 1$. Suppose you perform a measurement that identifies the complete set of states $|a_n\rangle$, that is

$$\sum_n |a_n\rangle\langle a_n| = 1\,. \tag{1.6.32}$$

Then the probabilities of the various outcomes are $\langle a_n|\rho|a_n\rangle$ which are assuredly positive, and their sum must be 1:

$$\begin{aligned} 1 &= \sum_n \langle a_n|\rho|a_n\rangle = \sum_n \mathrm{tr}\{\rho|a_n\rangle\langle a_n|\} \\ &= \mathrm{tr}\Bigl\{\rho \underbrace{\sum_n |a_n\rangle\langle a_n|}_{=1}\Bigr\} = \mathrm{tr}\{\rho\}\,. \end{aligned} \tag{1.6.33}$$

That is: $\mathrm{tr}\{\rho\} = 1$ is just the statement that the probabilities of mutually exclusive events have unit sum.

The extreme situation of only one term is the one we started with, $\rho = |\;\rangle\langle\;|$. Then it is possible to almost identify the ingredients. "Almost" because of the phase arbitrariness,

$$\rho = |\;\rangle\langle\;| = \Bigl(|\;\rangle e^{i\varphi}\Bigr)\Bigl(e^{-i\varphi}\langle\;|\Bigr)\,, \tag{1.6.34}$$

according to which the pair

$$|\;\rangle e^{i\varphi}\,, \quad e^{-i\varphi}\langle\;| \qquad (\varphi \text{ real}) \tag{1.6.35}$$

is as good as the pair $|\;\rangle$, $\langle\;|$. It is one of the advantages of using the statistical operator rather than $|\;\rangle$ and $\langle\;|$ that there is no phase arbitrariness in ρ. The statistical operator ρ is unique, its ingredients are not.

The situation of $\rho = |\;\rangle\langle\;|$ is also special because, for such a *pure* state, it is characteristically true that $\rho^2 = \rho$; see

$$\begin{aligned} \rho^2 &= \Bigl(|\;\rangle\langle\;|\Bigr)\Bigl(|\;\rangle\langle\;|\Bigr) = |\;\rangle\underbrace{\langle\;|\;\rangle}_{=1}\langle\;| \\ &= |\;\rangle\langle\;| = \rho\,. \end{aligned} \tag{1.6.36}$$

If there is more than one term in $\rho = \sum_k |k\rangle w_k \langle k|$ then $\rho^2 \neq \rho$, as is best illustrated by considering the trace

$$\begin{aligned} \operatorname{tr}\{\rho^2\} &= \sum_{j,k} \operatorname{tr}\{|j\rangle w_j \langle j|k\rangle w_k \langle k|\} \\ &= \sum_{j,k} w_j w_k |\langle j|k\rangle|^2 . \end{aligned} \tag{1.6.37}$$

Now, since $|j\rangle\langle j| \neq |k\rangle\langle k|$ if $j \neq k$ (we want *really* different ingredients), we have $|\langle j|k\rangle|^2 < 1$ for $j \neq k$, so that

$$\operatorname{tr}\{\rho^2\} < \sum_{j,k} w_j w_k = \underbrace{\left(\sum_j w_j\right)}_{=1} \underbrace{\left(\sum_k w_k\right)}_{=1} = 1 , \tag{1.6.38}$$

or $\operatorname{tr}\{\rho^2\} < 1$. Thus we have

$$\begin{aligned} &\operatorname{tr}\{\rho^2\} = 1 \quad \text{if} \quad \rho = |\ \rangle\langle\ |, \quad \rho^2 = \rho \\ \text{and} \quad &\operatorname{tr}\{\rho^2\} < 1 \quad \text{otherwise}. \end{aligned} \tag{1.6.39}$$

Therefore, the number $\operatorname{tr}\{\rho^2\}$ can serve as a crude measure of the purity of the state, it is maximal, $\operatorname{tr}\{\rho^2\} = 1$, for a pure state, $\rho = |\ \rangle\langle\ |$, and surely less than unity for all truly mixed states.

1.7 Algebraic completeness of operators X and P

We have the position operator X and the momentum operator P, functions $f(X)$, $g(P)$ of either one, and upon forming products and sums of such functions can introduce rather arbitrary functions of both X and P. And these general functions $f(X,P)$ comprise *all* possible operators for a degree of freedom of this sort. In other words: position X and momentum P are *algebraically complete.*

To demonstrate this we show that we can write any given operator A as a function of X and P, quite systematically. We begin with noting that $\langle x|A$ is a well defined bra and $A|p\rangle$ is a well defined ket, and that $\langle x|A|p\rangle$ is a uniquely specified set of numbers once A is stated. These numbers appear in

$$A = 1\,A\,1 = \int \mathrm{d}x\,|x\rangle\langle x|A\int \mathrm{d}p\,|p\rangle\langle p| \\ = \int \mathrm{d}x\,\mathrm{d}p\,|x\rangle\langle x|A|p\rangle\langle p|\,. \tag{1.7.1}$$

We divide and multiply by

$$\langle x|p\rangle = \frac{1}{\sqrt{2\pi\hbar}}\,\mathrm{e}^{\mathrm{i}xp/\hbar}\,, \tag{1.7.2}$$

which is never zero, to arrive at

$$A = \int \mathrm{d}x\,\mathrm{d}p\,|x\rangle\langle x|a(x,p)|p\rangle\langle p| \tag{1.7.3}$$

where

$$a(x,p) = \frac{\langle x|A|p\rangle}{\langle x|p\rangle} \tag{1.7.4}$$

is such that $A = 1$ is mapped onto $a(x,p) = 1$. Borrowing once again the terminology from classical mechanics, we call $a(x,p)$ a *phase-space function* of A.

1-14 Show that this mapping $A \to a(x,p)$ is linear. What is $a(x,p)$ for $A = f(X)$? For $A = g(P)$?

Further, consistent with the general rules

$$f(X) = \int \mathrm{d}x'\,|x'\rangle f(x')\langle x'|\,, \\ g(P) = \int \mathrm{d}p'\,|p'\rangle g(p')\langle p'| \tag{1.7.5}$$

we have

$$|x\rangle\langle x| = \int \mathrm{d}x'\,|x'\rangle\delta(x'-x)\langle x'| = \delta(X-x)\,, \\ |p\rangle\langle p| = \int \mathrm{d}p'\,|p'\rangle\delta(p'-p)\langle p'| = \delta(P-p) \tag{1.7.6}$$

and therefore

$$A = \int \mathrm{d}x\,\mathrm{d}p\;\delta(X-x)a(x,p)\delta(P-p)\,. \tag{1.7.7}$$

This equation already proves the case: we have expressed A as a function of X and P. But we can go one step further and evaluate the integrals over the δ functions, with the outcome

$$A = a(X,P)\Big|_{X,P\text{-ordered}} \equiv a(X;P) \tag{1.7.8}$$

where we must pay due attention to the structure of the previous expression (1.7.7). There all Xs stand to the left of all Ps, and this order must be preserved when we replace $x \to X$, $p \to P$ in $a(x,p)$.

We have thus achieved even more than what we really needed. Operator A is now expressed as an *ordered* function of X and P, for which the procedure gives a unique answer. Of course, we can interchange the roles of position and momentum in this argument and can equally well arrive at a unique P,X-ordered form, where all P operators are to the left of all X operators in products.

As a simple example, consider $A = PX$ for which

$$\begin{aligned}\langle x|A|p\rangle &= \langle x|PX|p\rangle \\ &= \frac{\hbar}{\mathrm{i}}\frac{\partial}{\partial x}\langle x|\frac{\hbar}{\mathrm{i}}\frac{\partial}{\partial p}|p\rangle \\ &= \left(\frac{\hbar}{\mathrm{i}}\right)^2\frac{\partial}{\partial x}\frac{\partial}{\partial p}\langle x|p\rangle \\ &= \left(\frac{\hbar}{\mathrm{i}}\right)^2\left(\frac{\mathrm{i}}{\hbar}+\left(\frac{\mathrm{i}}{\hbar}\right)^2 xp\right)\langle x|p\rangle\,,\end{aligned} \tag{1.7.9}$$

where the last step exploits the familiar explicit form of $\langle x|p\rangle$ in (1.7.2). Accordingly,

$$a(x,p) = \frac{\langle x|A|p\rangle}{\langle x|p\rangle} = xp + \frac{\hbar}{\mathrm{i}} \tag{1.7.10}$$

here, and we get the X,P-ordered form

$$A = PX = \left(xp + \frac{\hbar}{\mathrm{i}}\right)\Bigg|_{\substack{x\,\to\, X\\ p\,\to\, P\\ \text{ordered}}} = XP - \mathrm{i}\hbar\,. \tag{1.7.11}$$

The result is, of course, as expected inasmuch as we just get Werner Heisenberg's fundamental *commutation relation*

$$[X,P] = \mathrm{i}\hbar\,. \tag{1.7.12}$$

(See Section 3.2 in *Basic Matters* for the basic properties of commutators, in particular their linearity, expressed by the sum rule, and their product rule.) We recall the two extensions,

$$[f(X), P] = \mathrm{i}\hbar \frac{\partial f(X)}{\partial X}\,, \qquad [X, g(P)] = \mathrm{i}\hbar \frac{\partial g(P)}{\partial P}\,, \tag{1.7.13}$$

which are frequently used (see Section 5.1.4 in *Basic Matters*).

1-15 Show that, most generally,

$$[F(X,P), P] = \mathrm{i}\hbar \frac{\partial F(X,P)}{\partial X}\,, \qquad [X, F(X,P)] = \mathrm{i}\hbar \frac{\partial F(X,P)}{\partial P}\,,$$

where $F(X,P)$ is *any* operator function of X and P. It is sufficient (why?) to consider the special case $F = |x\rangle\langle p|$.

1-16 Is there a difference between $\frac{\partial}{\partial X}\frac{\partial}{\partial P} f(X,P)$ and $\frac{\partial}{\partial P}\frac{\partial}{\partial X} f(X,P)$?

1-17 Find the X,P-ordered form of the commutator $\frac{1}{\mathrm{i}\hbar}[X^2, P^2]$.

A simple statistical operator for a pure state, $\rho = |\;\rangle\langle\;|$, is our next example. We have

$$\langle x|\rho|p\rangle = \langle x|\;\rangle\langle\;|p\rangle = \psi(x)\psi(p)^* \tag{1.7.14}$$

and then

$$\frac{\langle x|\rho|p\rangle}{\langle x|p\rangle} = \frac{\langle x|\;\rangle\langle\;|p\rangle}{\langle x|p\rangle} = \sqrt{2\pi\hbar}\,\psi(x)\,\mathrm{e}^{-\mathrm{i}xp/\hbar}\psi(p)^* \tag{1.7.15}$$

so that

$$\rho = \sqrt{2\pi\hbar}\,\psi(X)\,\mathrm{e}^{-\mathrm{i}X;\,P/\hbar}\psi(P)^\dagger\,. \tag{1.7.16}$$

Here

$$\begin{aligned}\mathrm{e}^{-\mathrm{i}X;\,P/\hbar} &= \mathrm{e}^{-\mathrm{i}XP/\hbar}\Big|_{X,P\text{-ordered}} \\ &= \sum_{k=0}^{\infty}\frac{1}{k!}\left(\frac{-\mathrm{i}}{\hbar}\right)^k X^k P^k \end{aligned}\tag{1.7.17}$$

is a basic *ordered exponential function.* Its adjoint is

$$\left(\mathrm{e}^{-\mathrm{i}X;\,P/\hbar}\right)^\dagger = \mathrm{e}^{\mathrm{i}P;\,X/\hbar} \tag{1.7.18}$$

as one verifies immediately, and since $\rho^\dagger = \rho$ we get

$$\rho = \sqrt{2\pi\hbar}\,\psi(P)\,\mathrm{e}^{\mathrm{i}P;\,X/\hbar}\psi(X)^\dagger \tag{1.7.19}$$

for the P, X-ordered version of ρ.

When the ordered form of an operator is at hand, it is particularly easy to evaluate its trace,

$$\begin{aligned}\mathrm{tr}\{A\} &= \mathrm{tr}\left\{\int \mathrm{d}x\,|x\rangle\langle x|A\int \mathrm{d}p\,|p\rangle\langle p|\right\} \\ &= \int \mathrm{d}x\,\mathrm{d}p\,\langle x|A|p\rangle\langle p|x\rangle \\ &= \int \mathrm{d}x\,\mathrm{d}p\,\underbrace{\frac{\langle x|A|p\rangle}{\langle x|p\rangle}}_{a(x,p)=}\underbrace{\langle x|p\rangle\langle p|x\rangle}_{=1/(2\pi\hbar)}\end{aligned} \tag{1.7.20}$$

so that

$$\mathrm{tr}\{A\} = \int \frac{\mathrm{d}x\,\mathrm{d}p}{2\pi\hbar}\,a(x,p)\,. \tag{1.7.21}$$

This has the appearance of a classical phase-space integral, counting one quantum state per phase-space area of $2\pi\hbar$, so to say.

Next, suppose operator A is given in its X, P-ordered form,

$$A = a(X;P)\,, \tag{1.7.22}$$

and the statistical operator is given as a P, X-ordered expression

$$\rho = r(P;X)\,. \tag{1.7.23}$$

Then we have

$$\begin{aligned}\langle A\rangle &= \mathrm{tr}\{\rho A\} \\ &= \mathrm{tr}\left\{\int \mathrm{d}p\,|p\rangle\langle p|\rho\int \mathrm{d}x\,|x\rangle\langle x|A\right\} \\ &= \int \mathrm{d}x\,\mathrm{d}p\,\langle p|\rho|x\rangle\langle x|A|p\rangle\,,\end{aligned} \tag{1.7.24}$$

for the expectation value of A, where we meet

$$\langle p|\rho|x\rangle = \langle p|r(\underset{\substack{\downarrow\\ p}}{P};\underset{\substack{\downarrow\\ x}}{X})|x\rangle = r(p,x)\langle p|x\rangle \tag{1.7.25}$$

and

$$\langle x|A|p\rangle = \langle x|a(\underset{\substack{\downarrow\\ x}}{X};\underset{\substack{\downarrow\\ p}}{P})|p\rangle = a(x,p)\langle x|p\rangle\,. \tag{1.7.26}$$

These just exploit the basic property of ordered operators, namely that X and P stand next to their respective eigenbras and eigenkets so that these operators can be equivalently replaced by their eigenvalues. Then,

$$\langle A\rangle = \int \mathrm{d}x\,\mathrm{d}p\; r(p,x)a(x,p)\underbrace{\langle p|x\rangle\langle x|p\rangle}_{=1/(2\pi\hbar)} \tag{1.7.27}$$

or

$$\langle A\rangle = \int \frac{\mathrm{d}x\,\mathrm{d}p}{2\pi\hbar}\, r(p,x)a(x,p)\,, \tag{1.7.28}$$

which looks even more like a classical phase-space integral of the product of a density $r(p,x)$ and a phase-space function $a(x,p)$, whereby $\dfrac{\mathrm{d}x\,\mathrm{d}p}{2\pi\hbar}$ suggests the injunction noted at (1.7.21), namely to count one quantum state per phase-space area of $2\pi\hbar$.

The seemingly classical appearance of (1.7.28) is striking, it is also profound, but we must keep in mind that we continue to talk about quantum mechanical traces and that the phase-space functions $r(p,x)$ and $a(x,p)$ are just particularly convenient numerical descriptions of the quantum mechanical operators ρ and A. We are *not* replacing quantum mechanics by some equivalent version of classical mechanics. There is no such thing.

1-18 Use the X,P-ordered and P,X-ordered forms of $\rho = |\;\rangle\langle\;|$ to evaluate $\mathrm{tr}\{\rho\}$ as a phase-space integral.

1.8 Weyl commutator, Baker–Campbell–Hausdorff relations

The unitary operators $\mathrm{e}^{\mathrm{i}p'X/\hbar}$ and $\mathrm{e}^{\mathrm{i}x'P/\hbar}$ of (1.5.14) do not commute, except for special values of x' and p', and we can find out what is the difference between applying them in either order by establishing the X,P-ordered version of the operator

$$A = \mathrm{e}^{\mathrm{i}x'P/\hbar}\,\mathrm{e}^{\mathrm{i}p'X/\hbar}\,, \tag{1.8.1}$$

which we thus define by its P, X-ordered version. We begin with

$$\begin{aligned}\langle x|A|p\rangle &= \underbrace{\langle x|\,\mathrm{e}^{\mathrm{i}x'P/\hbar}}_{\langle x+x'|=}\underbrace{\mathrm{e}^{\mathrm{i}p'X/\hbar}|p\rangle}_{=|p+p'\rangle}\\ &= \langle x+x'|p+p'\rangle = \frac{1}{\sqrt{2\pi\hbar}}\,\mathrm{e}^{\mathrm{i}(x+x')(p+p')/\hbar}\\ &= \underbrace{\frac{\mathrm{e}^{\mathrm{i}xp/\hbar}}{\sqrt{2\pi\hbar}}}_{=\langle x|p\rangle}\underbrace{\mathrm{e}^{\mathrm{i}xp'/\hbar}\,\mathrm{e}^{\mathrm{i}x'p/\hbar}\,\mathrm{e}^{\mathrm{i}x'p'/\hbar}}_{=a(x,p)}\,,\end{aligned} \tag{1.8.2}$$

and now the replacement $a(x,p) \to a(X;P)$ gives us

$$A = \mathrm{e}^{\mathrm{i}p'X/\hbar}\,\mathrm{e}^{\mathrm{i}x'P/\hbar}\,\mathrm{e}^{\mathrm{i}x'p'/\hbar}\,. \tag{1.8.3}$$

More explicitly, this says

$$\mathrm{e}^{\mathrm{i}p'X/\hbar}\,\mathrm{e}^{\mathrm{i}x'P/\hbar} = \mathrm{e}^{-\mathrm{i}x'p'/\hbar}\,\mathrm{e}^{\mathrm{i}x'P/\hbar}\,\mathrm{e}^{\mathrm{i}p'X/\hbar}\,, \tag{1.8.4}$$

which is Hermann K. H. Weyl's commutation relation for the basic unitary operators associated with X and P, the *Weyl commutator* for short.

How about combining the various exponentials into one? That requires some care because the arguments of the exponentials do not commute with each other. But nevertheless we can be systematic about it, and we could use several methods for this purpose. Let us do it with a sequence of unitary transformations.

As a preparation, we recall that

$$U^\dagger f(A)U = f(U^\dagger AU) \tag{1.8.5}$$

for any function of operator A (arbitrary) and any arbitrary operator U. One easily verifies that it is true for any power of A — for example

$$U^\dagger A^2 U = U^\dagger AUU^\dagger AU = \left(U^\dagger AU\right)^2 \tag{1.8.6}$$

— and then it is true for all polynomials and finally for arbitrary functions.

1-19 If $|a\rangle$ is an eigenket of A, $A|a\rangle = |a\rangle a$, then $f(A)|a\rangle = |a\rangle f(a)$. Why? Conclude that $U^\dagger|a\rangle$ is an eigenket of $f(U^\dagger AU)$.

We also note that the differentiation rules in (1.7.13) imply

$$\begin{aligned}\left[X,\, \mathrm{e}^{\mathrm{i}g(P)}\right] &= -\hbar g'(P)\,\mathrm{e}^{\mathrm{i}g(P)}\,,\\ \left[\mathrm{e}^{\mathrm{i}f(X)},\, P\right] &= -\hbar f'(X)\,\mathrm{e}^{\mathrm{i}f(X)}\,,\end{aligned} \tag{1.8.7}$$

which are equivalent to

$$\begin{aligned}\mathrm{e}^{-\mathrm{i}g(P)}X\,\mathrm{e}^{\mathrm{i}g(P)} &= X - \hbar g'(P)\,,\\ \mathrm{e}^{\mathrm{i}f(X)}P\,\mathrm{e}^{-\mathrm{i}f(X)} &= P - \hbar f'(X)\,,\end{aligned} \tag{1.8.8}$$

with primes indicating differentiation with respect to the argument.

In a first step we write

$$\mathrm{e}^{\mathrm{i}(p'X + x'P)/\hbar} = \mathrm{e}^{\mathrm{i}p'(X + \frac{x'}{p'}P)/\hbar} \quad \text{for} \quad p' \neq 0 \tag{1.8.9}$$

and note that

$$X + \frac{x'}{p'}P = \mathrm{e}^{\frac{\mathrm{i}}{2\hbar}\frac{x'}{p'}P^2}\,X\,\mathrm{e}^{-\frac{\mathrm{i}}{2\hbar}\frac{x'}{p'}P^2} \tag{1.8.10}$$

and therefore

$$\begin{aligned}\mathrm{e}^{\mathrm{i}(p'X + x'P)/\hbar} &= \exp\left(\mathrm{i}p'\,\mathrm{e}^{\frac{\mathrm{i}}{2\hbar}\frac{x'}{p'}P^2}\,X\,\mathrm{e}^{-\frac{\mathrm{i}}{2\hbar}\frac{x'}{p'}P^2}/\hbar\right)\\ &= \mathrm{e}^{\frac{\mathrm{i}}{2\hbar}\frac{x'}{p'}P^2}\,\mathrm{e}^{\mathrm{i}p'X/\hbar}\,\mathrm{e}^{-\frac{\mathrm{i}}{2\hbar}\frac{x'}{p'}P^2}\,.\end{aligned} \tag{1.8.11}$$

Now we remember that we want to have a factor $\mathrm{e}^{\mathrm{i}p'X/\hbar}$ on the right eventually, so we write

$$\mathrm{e}^{\mathrm{i}(p'X + x'P)/\hbar} = \mathrm{e}^{\frac{\mathrm{i}}{2\hbar}\frac{x'}{p'}P^2}\,\mathrm{e}^{\mathrm{i}p'X/\hbar}\,\mathrm{e}^{-\frac{\mathrm{i}}{2\hbar}\frac{x'}{p'}P^2}\,\mathrm{e}^{-\mathrm{i}p'X/\hbar}\,\mathrm{e}^{\mathrm{i}p'X/\hbar} \tag{1.8.12}$$

and observe that

$$\mathrm{e}^{\mathrm{i}p'X/\hbar}\,\mathrm{e}^{-\frac{\mathrm{i}}{2\hbar}\frac{x'}{p'}P^2}\,\mathrm{e}^{-\mathrm{i}p'X/\hbar} = \exp\left(-\frac{\mathrm{i}}{2\hbar}\frac{x'}{p'}\left(\mathrm{e}^{\mathrm{i}p'X/\hbar}P\,\mathrm{e}^{-\mathrm{i}p'X/\hbar}\right)^2\right) \tag{1.8.13}$$

wherein

$$\mathrm{e}^{\mathrm{i}p'X/\hbar}P\,\mathrm{e}^{-\mathrm{i}p'X/\hbar} = P - p'\,, \tag{1.8.14}$$

implying

$$\mathrm{e}^{\mathrm{i}(p'X + x'P)/\hbar} = \mathrm{e}^{\frac{\mathrm{i}}{2\hbar}\frac{x'}{p'}P^2}\,\mathrm{e}^{-\frac{\mathrm{i}}{2\hbar}\frac{x'}{p'}(P-p')^2}\,\mathrm{e}^{\mathrm{i}p'X/\hbar}\,. \tag{1.8.15}$$

The first and second exponentials on the right are functions of P only and so there is no problem in combining them into one,

$$\begin{aligned} e^{\frac{i}{2\hbar}\frac{x'}{p'}P^2}\,e^{-\frac{i}{2\hbar}\frac{x'}{p'}(P-p')^2} &= e^{\frac{i}{2\hbar}\frac{x'}{p'}\left[P^2-(P^2-p')^2\right]} \\ &= e^{\frac{i}{2\hbar}\frac{x'}{p'}(2P-p')p'} \\ &= e^{ix'P/\hbar}\,e^{-\frac{i}{2}x'p'/\hbar}\,. \end{aligned} \tag{1.8.16}$$

Accordingly,

$$\begin{aligned} e^{i(p'X+x'P)/\hbar} &= e^{ix'P/\hbar}\,e^{ip'X/\hbar}\,e^{-\frac{i}{2}x'p'/\hbar} \\ &= e^{ip'X/\hbar}\,e^{ix'P/\hbar}\,e^{\frac{i}{2}x'p'/\hbar} \end{aligned} \tag{1.8.17}$$

where the second equality is that of the right-hand sides in (1.8.1) and (1.8.2). These are examples of the famous *Baker–Campbell–Hausdorff relations* among exponential functions of operators, named after Henry F. Baker, John E. Campbell, and Felix Hausdorff.

1-20 What is $\langle x|\,e^{i(p'X+x'P)/\hbar}$? What is $e^{i(p'X+x'P)/\hbar}|p\rangle$? Introduce the operator

$$R=\int\frac{dx'\,dp'}{2\pi\hbar}\,e^{i(p'X+x'P)/\hbar}$$

and find $\langle x|R$ and $R|p\rangle$.

1-21 Show that

$$\mathrm{tr}\left\{e^{i(p'X+x'P)/\hbar}e^{-i(p''X+x''P)/\hbar}\right\}=2\pi\hbar\delta(x'-x'')\delta(p'-p'')\,.$$

1-22 For any (reasonable) operator function $A(X,P)$ we can define its *characteristic function*

$$a(x,p)=\mathrm{tr}\left\{A(X,P)\,e^{i(xP+pX)/\hbar}\right\}.$$

This is a mapping of operator $A(X,P)$ on the phase-space function $a(x,p)$. [The present $a(x,p)$ is *not* the same as the one in (1.7.4).] Show that the inverse map is given by

$$A(X,P)=\int\frac{dx\,dp}{2\pi\hbar}\,e^{i(xP+pX)/\hbar}a(x,p)\,.$$

Hint: It is enough to show this for all ket-bras of the form $|x'\rangle\langle p'|$. Why?

1-23 Find $a(x,p)$ for $A(X,P)=X^n$, $n=0,1,2,\ldots$.

Chapter 2

Quantum Dynamics Reviewed

2.1 Temporal evolution

Relations such as

$$\langle x|X = x\langle x|\,,\quad X|x\rangle = |x\rangle x\,,\quad \langle x|P = \frac{\hbar}{\mathrm{i}}\frac{\partial}{\partial x}\langle x|\,,$$
$$\text{or}\quad \langle x|p\rangle = \frac{1}{\sqrt{2\pi\hbar}}\,\mathrm{e}^{\mathrm{i}xp/\hbar}\,,\quad \int \mathrm{d}x\,|x\rangle\langle x| = 1 \tag{2.1.1}$$

and so forth, all refer implicitly to a particular instant in time, the moment at which we measure position, or momentum, or some other property represented by a function $f(X,P)$ of position X and momentum P. More generally, however, there is the option of measuring one property now and another earlier or later. Therefore, we must extend the formalism so that we can consistently deal with time-dependent quantities. In short, we must be able to handle temporal evolution.

First of all, let us note that a measurement of position, say, at two different times are two different measurements, and so we need a symbol $X(t_1)$ for the position measurement at time t_1, and a symbol $X(t_2)$ for measurement at time t_2. Quite generally, then, we have a symbol $X(t)$ for position measurements at time t. And going with it are kets $|x,t\rangle$ and bras $\langle x,t|$ that refer to time t. The eigenket equation and the eigenbra equation above then generalize to

$$X(t)|x,t\rangle = |x,t\rangle x\,,\qquad \langle x,t|X(t) = x\langle x,t|\,. \tag{2.1.2}$$

Note that the eigenvalue x does not depend on time t, because it is a stand-in for all possible measurement results, and they are the same at all times t. The orthonormality and completeness relations for the time-dependent kets and bras,

$$\langle x,t|x',t\rangle = \delta(x-x'),$$
$$\int \mathrm{d}x\,|x,t\rangle\langle x,t| = 1, \tag{2.1.3}$$

are of the same appearance as the time-independent ones in (2.1.1). After all, we are just making the implicit time label explicit.

This story repeats for momentum P, where we have

$$P(t)|p,t\rangle = |p,t\rangle p, \qquad \langle p,t|P(t) = p\langle p,t|,$$
$$\langle p,t|p',t\rangle = \delta(p-p'),$$
$$\int \mathrm{d}p\,|p,t\rangle\langle p,t| = 1, \tag{2.1.4}$$

and the links between the position description and the momentum description are also just as before:

$$\langle x,t|p,t\rangle = \frac{1}{\sqrt{2\pi\hbar}}\,\mathrm{e}^{\mathrm{i}xp/\hbar},$$
$$\langle x,t|P(t) = \frac{\hbar}{\mathrm{i}}\frac{\partial}{\partial x}\langle x,t|, \qquad \langle p,t|X(t) = \mathrm{i}\hbar\frac{\partial}{\partial p}\langle p,t|. \tag{2.1.5}$$

In general terms: All relations that we had so far remain true as long as all kets, all bras, and all operators refer to a common time. This includes the Heisenberg commutation relation

$$[X(t),P(t)] = \mathrm{i}\hbar \tag{2.1.6}$$

in particular, and also the Baker–Campbell–Hausdorff relations of (1.8.17).

How do we relate the description at one time to that at another time? Clearly, a mapping of the kets and bras at time t, to those at time $t+\tau$,

$$\langle x,t| \to \langle x,t+\tau| = \langle x,t|U,$$
$$|p,t\rangle \to |p,t+\tau\rangle = U^{\dagger}|p,t\rangle, \tag{2.1.7}$$

must be a unitary transformation in order to preserve all geometrical relations among the kets and bras. Consider in particular the transformation function,

$$\frac{1}{\sqrt{2\pi\hbar}}\,\mathrm{e}^{\mathrm{i}xp/\hbar} = \langle x,t|p,t\rangle = \langle x,t+\tau|p,t+\tau\rangle = \langle x,t|UU^\dagger|p,t\rangle\,, \tag{2.1.8}$$

which requires

$$\langle x,t|p,t\rangle = \langle x,t|UU^\dagger|p,t\rangle \tag{2.1.9}$$

for all quantum numbers x and p. The completeness of the x states and the p states then implies immediately that the operator U in (2.1.7) is unitary,

$$UU^\dagger = 1\,. \tag{2.1.10}$$

We go from time t to a later time in a succession of small steps, eventually in a succession of infinitesimal steps. So, let us take the increment τ to be infinitesimal. Then the *unitary evolution operator* U will differ from the identity operator by an infinitesimal amount proportional to τ,

$$U = 1 - \frac{\mathrm{i}}{\hbar}H\tau\,, \tag{2.1.11}$$

evolution operator for an infinitesimal time step (U); infinitesimal time step (τ); generator for time changes (H)

which you may regard as the definition of H, the operator that generates changes in time. We borrow the terminology from classical physics, where the corresponding object is the Hamilton function, named after William R. Hamilton, and so we call H the *Hamilton operator.*

The factor $\frac{\mathrm{i}}{\hbar}$ is included in this definition for a double purpose. The i ensures that H is hermitian, $H = H^\dagger$, see

$$1 = UU^\dagger = \left(1 - \frac{\mathrm{i}}{\hbar}H\tau\right)\left(1 + \frac{\mathrm{i}}{\hbar}H^\dagger\tau\right) = 1 - \frac{\mathrm{i}}{\hbar}(H - H^\dagger)\tau, \quad \text{so that } H = H^\dagger\,. \tag{2.1.12}$$

And the $\hbar$ gives H the metrical dimension of energy, because τ is a time and $\hbar$ has the metrical dimension of energy $\times$ time.

Now, from

$$\begin{aligned}\langle x,t+\tau| &= \langle x,t|\left(1-\frac{\mathrm{i}}{\hbar}H\tau\right),\\ |p,t+\tau\rangle &= \left(1+\frac{\mathrm{i}}{\hbar}H\tau\right)|p,t\rangle\,,\end{aligned}\tag{2.1.13}$$

we get the differential statements

$$\begin{aligned}\frac{\partial}{\partial t}\langle x,t| &= \frac{1}{\tau}\Big(\langle x,t+\tau|-\langle x,t|\Big)_{\tau\to 0}\\ &= \langle x,t|\left(-\frac{\mathrm{i}}{\hbar}H\right)\end{aligned}\tag{2.1.14}$$

and

$$\begin{aligned}\frac{\partial}{\partial t}|p,t\rangle &= \frac{1}{\tau}\Big(|p,t+\tau\rangle-|p,t\rangle\Big)_{\tau\to 0}\\ &= \frac{\mathrm{i}}{\hbar}H|p,t\rangle\end{aligned}\tag{2.1.15}$$

or

$$\begin{aligned}\mathrm{i}\hbar\frac{\partial}{\partial t}\langle x,t| &= \langle x,t|H\,,\\ -\mathrm{i}\hbar\frac{\partial}{\partial t}|p,t\rangle &= H|p,t\rangle\,,\end{aligned}\tag{2.1.16}$$

and the adjoint statements read

$$\begin{aligned}-\mathrm{i}\hbar\frac{\partial}{\partial t}|x,t\rangle &= H|x,t\rangle\,,\\ \mathrm{i}\hbar\frac{\partial}{\partial t}\langle p,t| &= \langle p,t|H\,.\end{aligned}\tag{2.1.17}$$

The two bra equations are obviously particular examples of

$$\mathrm{i}\hbar\frac{\partial}{\partial t}\langle\ldots,t| = \langle\ldots,t|H\tag{2.1.18}$$

with the ellipsis standing for any corresponding label, any set of quantum numbers specifying the bra. This is the general form of Erwin Schrödinger's equation of motion — the celebrated *Schrödinger equation*, here for bras. The adjoint statement

$$-\mathrm{i}\hbar\frac{\partial}{\partial t}|\ldots,t\rangle = H|\ldots,t\rangle\tag{2.1.19}$$

is the Schrödinger equation for kets, exemplified by the two ket equations in (2.1.16) and (2.1.17).

We recall that in the present context of motion along the x axis all operators are functions of X and P, now more precisely: of $X(t)$ and $P(t)$. This remark applies in particular to the Hamilton operator,

$$H = H\bigl(X(t), P(t), t\bigr)\,, \tag{2.1.20}$$

where we note the possibility of a parametric time dependence as well, that is: at different times we may have structurally different functions of X and P for the Hamilton operator.

Upon invoking the differential-operator representation for P in (2.1.5), we have

$$\bigl\langle x,t\bigr|H\bigl(X(t),P(t),t\bigr) = H\left(x,\frac{\hbar}{\mathrm{i}}\frac{\partial}{\partial x},t\right)\bigl\langle x,t\bigr| \tag{2.1.21}$$

so that the position wave function to ket $|\ \rangle$,

$$\psi(x,t) = \bigl\langle x,t\bigr|\ \bigr\rangle\,, \tag{2.1.22}$$

obeys the differential equation

$$\mathrm{i}\hbar\frac{\partial}{\partial t}\psi(x,t) = H\left(x,\frac{\hbar}{\mathrm{i}}\frac{\partial}{\partial x},t\right)\psi(x,t)\,. \tag{2.1.23}$$

This is often referred to as *the* Schrödinger equation, and equally frequently one meets this name association for the typical case of

$$H = \underbrace{\frac{1}{2M}P^2}_{\text{kinetic energy}} + \underbrace{V(X)}_{\text{potential energy}} \tag{2.1.24}$$

when

$$\mathrm{i}\hbar\frac{\partial}{\partial t}\psi(x,t) = \left(-\frac{\hbar^2}{2M}\frac{\partial^2}{\partial x^2} + V(x)\right)\psi(x,t)\,. \tag{2.1.25}$$

These are, however, nothing more than quite special versions of (2.1.18), special versions of great practical importance, yes, but special versions nevertheless.

The evolution of operator $X(t)$ is now found from the evolution of the bras and kets in

$$X(t) = \int \mathrm{d}x\,\bigl|x,t\bigr\rangle x\bigl\langle x,t\bigr|\,, \tag{2.1.26}$$

namely

$$\begin{aligned}\frac{\mathrm{d}}{\mathrm{d}t}X(t) &= \int \mathrm{d}x \left(\frac{\partial|x,t\rangle}{\partial t}x\langle x,t| + |x,t\rangle x\frac{\partial\langle x,t|}{\partial t}\right)\\ &= \int \mathrm{d}x \left(\frac{\mathrm{i}}{\hbar}H|x,t\rangle x\langle x,t| + |x,t\rangle x\langle x,t|\frac{1}{\mathrm{i}\hbar}H\right)\\ &= \frac{\mathrm{i}}{\hbar}HX + \frac{1}{\mathrm{i}\hbar}XH = \frac{1}{\mathrm{i}\hbar}[X,H]\,,\end{aligned} \tag{2.1.27}$$

or more explicitly,

$$\frac{\mathrm{d}}{\mathrm{d}t}X(t) = \frac{1}{\mathrm{i}\hbar}\Big[X(t), H\big(X(t),P(t),t\big)\Big]\,. \tag{2.1.28}$$

The argument can be repeated, with the necessary changes, for $P(t)$ with the analogous outcome

$$\frac{\mathrm{d}}{\mathrm{d}t}P(t) = \frac{1}{\mathrm{i}\hbar}\Big[P(t), H\big(X(t),P(t),t\big)\Big]\,. \tag{2.1.29}$$

These are examples of Werner Heisenberg's equations of motion, the *Heisenberg equation.* We shall get to the general form shortly. Right now, however, we recall the lesson of Exercise 1-15 on page 35, namely that

$$[X,f] = \mathrm{i}\hbar\frac{\partial f}{\partial P} \quad \text{and} \quad [f,P] = \mathrm{i}\hbar\frac{\partial f}{\partial X} \tag{2.1.30}$$

for any operator function of X and P — or, more pedantically, of $X(t)$, $P(t)$, and possibly t as a parameter. So

$$\frac{\mathrm{d}}{\mathrm{d}t}X = \frac{\partial H}{\partial P}\,, \quad \frac{\mathrm{d}}{\mathrm{d}t}P = -\frac{\partial H}{\partial X}\,, \tag{2.1.31}$$

which have exactly the same appearance as William R. Hamilton's equations of motion in classical mechanics. These equations are correct on the classical level because they are already true on the quantum level.

What about the time derivative of an arbitrary operator function of X and P, $F = F(X,P,t)$? We know that we can always exploit the completeness of the x states and the p states to write it as

$$F = \int \mathrm{d}x\,\mathrm{d}p\,|x,t\rangle f(x,p,t)\langle p,t| \tag{2.1.32}$$

with

$$f(x,p,t) = \langle x,t|F\big(X(t),P(t),t\big)|p,t\rangle\,. \tag{2.1.33}$$

The time derivative of F has three contributions,

$$\frac{\mathrm{d}}{\mathrm{d}t}F = \int \mathrm{d}x\,\mathrm{d}p\,\underbrace{\frac{\partial|x,t\rangle}{\partial t}}_{=\frac{\mathrm{i}}{\hbar}H|x,t\rangle} f(x,p,t)\langle p,t| + \int \mathrm{d}x\,\mathrm{d}p\,|x,t\rangle f(x,p,t)\underbrace{\frac{\partial\langle p,t|}{\partial t}}_{=\langle p,t|\frac{1}{\mathrm{i}\hbar}H} + \int \mathrm{d}x\,\mathrm{d}p\,|x,t\rangle\frac{\partial f(x,p,t)}{\partial t}\langle p,t|\,, \tag{2.1.34}$$

of which the first is $\frac{\mathrm{i}}{\hbar}HF$, and the second is $\frac{1}{\mathrm{i}\hbar}FH$, so that they together are $\frac{1}{\mathrm{i}\hbar}[F,H]$. The third and last contribution is the parametric time derivative of F,

$$\frac{\partial}{\partial t}F\big(X(t),P(t),t\big) = \int \mathrm{d}x\,\mathrm{d}p\,|x,t\rangle\frac{\partial f(x,p,t)}{\partial t}\langle p,t|\,, \tag{2.1.35}$$

which we see after first noting that the time arguments in

$$f(x,p,t) = \langle x,t|F\big(X(t),P(t),t\big)|p,t\rangle \tag{2.1.36}$$

are of different significance. We demonstrate the nature of the arbitrary common time by labeling it as t' in

$$\begin{aligned}\langle x,t'|F\big(X(t'),P(t'),t\big)|p,t'\rangle &= F\left(x,\frac{\hbar}{\mathrm{i}}\frac{\partial}{\partial x},t\right)\langle x,t'|p,t'\rangle \\ &= F\left(x,\frac{\hbar}{\mathrm{i}}\frac{\partial}{\partial x},t\right)\frac{1}{\sqrt{2\pi\hbar}}\,\mathrm{e}^{\mathrm{i}xp/\hbar}\end{aligned} \tag{2.1.37}$$

where it is now obvious that this number is really independent of the common time t' to which X and P refer. As a consequence,

$$\frac{\partial}{\partial t}f(x,p,t) = \langle x,t|\frac{\partial F\big(X(t),P(t),t\big)}{\partial t}|p,t\rangle\,, \tag{2.1.38}$$

and the third term in $\frac{\mathrm{d}}{\mathrm{d}t}F$ is $\frac{\partial F}{\partial t}$, indeed.

In summary, then, we have

$$\frac{\mathrm{d}}{\mathrm{d}t}F = \underbrace{\frac{1}{\mathrm{i}\hbar}[F,H]}_{\substack{\text{dynamical}\\ \text{time dependence}}} + \underbrace{\frac{\partial}{\partial t}F}_{\substack{\text{parametric}\\ \text{time dependence}}} . \tag{2.1.39}$$

This is the general form of *Heisenberg's equation of motion.* We have thus

> the Schrödinger equation for the temporal evolution of bras, kets, and wave functions

and

> the Heisenberg equation for the temporal evolution of operators.

There are two special cases of the Heisenberg equation. First, if $F = X$ or $F = P$, then there is no time dependence and we have

$$\frac{\mathrm{d}}{\mathrm{d}t}X = \frac{1}{\mathrm{i}\hbar}[X,H]\,, \quad \frac{\mathrm{d}}{\mathrm{d}t}P = \frac{1}{\mathrm{i}\hbar}[P,H] \tag{2.1.40}$$

as above (without the $\frac{\partial}{\partial t}$ term). Second, if F is the statistical operator $\rho = |\ \rangle\langle\ |$, or a convex sum of such projection operators, then there is no total time dependence,

$$\frac{\mathrm{d}}{\mathrm{d}t}\rho = 0 \tag{2.1.41}$$

because, by its physical nature, ρ refers to a particular time at which the state of affairs is specified — such as: at $t = 0$ the system is described by the initial wave function $\psi_0(x)$ — and therefore does not depend on the evolution time t. For a statistical operator, we thus have

$$0 = \frac{\partial}{\partial t}\rho + \frac{1}{\mathrm{i}\hbar}[\rho,H] \tag{2.1.42}$$

or

$$\frac{\partial}{\partial t}\rho = \frac{\mathrm{i}}{\hbar}[\rho,H]\,. \tag{2.1.43}$$

This quantum analog of Joseph Liouville's equation of motion of classical statistical physics is known as the *von Neumann equation* (John von Neumann).

More explicitly the constancy in time of ρ means that

$$\rho = \rho\bigl(X(t_1),P(t_1),t_1\bigr) = \rho\bigl(X(t_2),P(t_2),t_2\bigr) \tag{2.1.44}$$

where we have different functions of X and P at the two times but such that, if the first function is taken for $X(t_1)$ and $P(t_1)$ and the second function for $X(t_2)$ and $P(t_2)$, we obtain the same operator ρ as a result. That is: the parametric time dependence compensates exactly for the dynamical time dependence. All of this will get clearer when we study some simple but instructive examples.

2-1 For the Hamilton operator of force-free motion,

$$H = \frac{1}{2M}P^2 ,$$

wherein the mass M does not vary in time, evaluate $\frac{\mathrm{d}}{\mathrm{d}t}F$ for $F = X$, $F = P$, and $F = MX - tP$.

2.2 Time transformation functions

Finding a solution of the Schrödinger equation means to express the wave function at a later time in terms of the given wave function at the earlier initial time. Using the completeness of the x states at the earlier time t_0, we can express the later wave function as

$$\begin{aligned}\psi(x,t) = \langle x,t|\ \rangle &= \int \mathrm{d}x' \,\langle x,t|x',t_0\rangle\langle x',t_0|\ \rangle \\ &= \int \mathrm{d}x' \,\langle x,t|x',t_0\rangle\, \psi(x',t_0)\,. \end{aligned} \tag{2.2.1}$$

Accordingly, the problem is reduced to finding the *time transformation function* $\langle x,t|x',t_0\rangle$ that relates the x description at t_0 to that at t. The Schrödinger equation for bra $\langle x,t|$ implies

$$\begin{aligned}\mathrm{i}\hbar\frac{\partial}{\partial t}\langle x,t|x',t_0\rangle &= \langle x,t|H\big(X(t),P(t),t\big)|x',t_0\rangle \\ &= H\Big(x,\frac{\hbar}{\mathrm{i}}\frac{\partial}{\partial x},t\Big)\langle x,t|x',t_0\rangle \end{aligned} \tag{2.2.2}$$

as the equation of motion for $\langle x,t|x',t_0\rangle$, to be solved with the initial condition

$$\langle x,t|x',t_0\rangle\Big|_{t\,\to\,t_0} = \delta(x-x')\,. \tag{2.2.3}$$

But this is only one of many possibilities. We could also relate the momentum descriptions to each other,

$$\psi(p,t) = \int \mathrm{d}p' \, \langle p,t|p',t_0\rangle \, \psi(p',t_0) \,, \tag{2.2.4}$$

where the time transformation function $\langle p,t|p',t_0\rangle$ obeys the Schrödinger equation

$$\begin{aligned} \mathrm{i}\hbar\frac{\partial}{\partial t}\langle p,t|p',t_0\rangle &= \langle p,t|H\bigl(X(t),P(t),t\bigr)|p',t_0\rangle \\ &= H\left(\mathrm{i}\hbar\frac{\partial}{\partial p},p,t\right)\langle p,t|p',t_0\rangle \,, \end{aligned} \tag{2.2.5}$$

subject to the initial condition

$$\langle p,t|p',t_0\rangle\Big|_{t\,\to\,t_0} = \delta(p-p') \,. \tag{2.2.6}$$

Or we relate $\langle x,t|$ to $|p,t_0\rangle$,

$$\psi(x,t) = \int \mathrm{d}p \, \langle x,t|p,t_0\rangle \, \psi(p,t_0) \,, \tag{2.2.7}$$

so that we turn the momentum wave function at t_0 into the position wave function at time t. Here we meet the xp time transformation function $\langle x,t|p,t_0\rangle$ for which

$$\mathrm{i}\hbar\frac{\partial}{\partial t}\langle x,t|p,t_0\rangle = H\left(x,\frac{\hbar}{\mathrm{i}}\frac{\partial}{\partial x},t\right)\langle x,t|p,t_0\rangle \tag{2.2.8}$$

is the Schrödinger equation, and

$$\langle x,t|p,t_0\rangle\Big|_{t\,\to\,t_0} = \frac{\mathrm{e}^{\mathrm{i}xp/\hbar}}{\sqrt{2\pi\hbar}} \tag{2.2.9}$$

is the initial condition.

Finally, we have the px time transformation function $\langle p,t|x,t_0\rangle$ in

$$\psi(p,t) = \int \mathrm{d}x \, \langle p,t|x,t_0\rangle \, \psi(x,t_0) \,. \tag{2.2.10}$$

Its Schrödinger equation reads

$$\mathrm{i}\hbar\frac{\partial}{\partial t}\langle p,t|x,t_0\rangle = H\left(\mathrm{i}\hbar\frac{\partial}{\partial p},p,t\right)\langle p,t|x,t_0\rangle \tag{2.2.11}$$

and the initial condition is

$$\left\langle p,t\middle|x,t_0\right\rangle\Big|_{t\,\to\,t_0} = \frac{\mathrm{e}^{-\mathrm{i}px/\hbar}}{\sqrt{2\pi\hbar}}\,. \tag{2.2.12}$$

As soon as we know one of the transformation functions, we can get the other ones by Fourier transformation, as illustrated by

$$\begin{aligned}\left\langle x,t\middle|x',t_0\right\rangle &= \int \mathrm{d}p\,\left\langle x,t\middle|p,t\right\rangle\left\langle p,t\middle|x',t_0\right\rangle \\ &= \int \mathrm{d}p\,\frac{\mathrm{e}^{\mathrm{i}xp/\hbar}}{\sqrt{2\pi\hbar}}\left\langle p,t\middle|x',t_0\right\rangle \\ &= \int \mathrm{d}p\,\left\langle x,t\middle|p,t_0\right\rangle\left\langle p,t_0\middle|x',t_0\right\rangle \\ &= \int \mathrm{d}p\,\left\langle x,t\middle|p,t_0\right\rangle\frac{\mathrm{e}^{-\mathrm{i}px'/\hbar}}{\sqrt{2\pi\hbar}}\,,\end{aligned} \tag{2.2.13}$$

where we get the xx version from the px or the xp forms. A bit more generally, but in the same spirit, are composition laws such as

$$\left\langle x,t\middle|x',t_0\right\rangle = \int \mathrm{d}p\,\left\langle x,t\middle|p,T\right\rangle\left\langle p,T\middle|x',t_0\right\rangle, \tag{2.2.14}$$

where T is any (intermediate) time, the extreme cases of $t = T$ and $T = t_0$ are those of above. This equation is actually an expression of the fact that evolution happens in steps: first you go from t_0 to T, then from T to t.

2-2 How can you get $\left\langle x,t\middle|p,t_0\right\rangle$ from $\left\langle p,t\middle|x,t_0\right\rangle$ without performing any Fourier transformations?

Chapter 3

Examples

3.1 Force-free motion

As a first example we consider force-free motion, for which the Hamilton operator is

$$H = \frac{1}{2M} P^2 , \tag{3.1.1}$$

just kinetic energy, no potential energy. The Heisenberg equations of motion,

$$\begin{aligned} \frac{\mathrm{d}}{\mathrm{d}t} P &= -\frac{\partial H}{\partial X} = 0 , \\ \frac{\mathrm{d}}{\mathrm{d}t} X &= \frac{\partial H}{\partial P} = \frac{1}{M} P , \end{aligned} \tag{3.1.2}$$

look exactly like their classical analogs, and we solve them easily,

$$\begin{aligned} P(t) &= P(t_0) , \\ X(t) &= X(t_0) + \frac{t - t_0}{M} P(t_0) . \end{aligned} \tag{3.1.3}$$

We put these aside for now and turn to the Schrödinger equation and the time transformation functions. Those with bra $\langle p, t|$ are simplest because for them, the Hamilton operator is a number,

$$\mathrm{i}\hbar \frac{\partial}{\partial t} \langle p, t| = \langle p, t| \frac{1}{2M} P(t)^2 = \frac{p^2}{2M} \langle p, t| , \tag{3.1.4}$$

so that

$$\begin{aligned}\langle p,t|x,t_0\rangle &= \mathrm{e}^{-\frac{\mathrm{i}}{\hbar}\frac{p^2}{2M}(t-t_0)}\langle p,t_0|x,t_0\rangle \\ &= \mathrm{e}^{-\frac{\mathrm{i}}{\hbar}\frac{p^2}{2M}(t-t_0)}\frac{\mathrm{e}^{-\mathrm{i}px/\hbar}}{\sqrt{2\pi\hbar}}\end{aligned} \tag{3.1.5}$$

is the immediate solution of the Schrödinger differential equation with the initial condition at $t = t_0$ correctly taken into account.

The xx time transformation function is now obtained by a Fourier transformation,

$$\begin{aligned}\langle x,t|x',t_0\rangle &= \int \mathrm{d}p\,\langle x,t|p,t\rangle\langle p,t|x',t_0\rangle \\ &= \int \mathrm{d}p\,\frac{\mathrm{e}^{\mathrm{i}xp/\hbar}}{\sqrt{2\pi\hbar}}\,\mathrm{e}^{-\frac{\mathrm{i}}{\hbar}\frac{p^2}{2M}(t-t_0)}\frac{\mathrm{e}^{-\mathrm{i}px/\hbar}}{\sqrt{2\pi\hbar}} \\ &= \int \frac{\mathrm{d}p}{2\pi\hbar}\,\mathrm{e}^{-\frac{\mathrm{i}}{\hbar}\frac{p^2}{2M}(t-t_0)}\,\mathrm{e}^{\mathrm{i}(x-x')p/\hbar}\,.\end{aligned} \tag{3.1.6}$$

This is a gaussian integral (Karl F. Gauss) which we evaluate in accordance with

$$\int \mathrm{d}y\;\mathrm{e}^{-\alpha y^2+\beta y} = \sqrt{\frac{\pi}{\alpha}}\,\mathrm{e}^{\frac{\beta^2}{4\alpha}} \quad \text{for} \quad \mathrm{Re}(\alpha)\geq 0\,, \tag{3.1.7}$$

here for

$$\alpha = \frac{\mathrm{i}}{\hbar}\frac{t-t_0}{2M} \quad \text{and} \quad \beta = \frac{\mathrm{i}}{\hbar}(x-x')\,. \tag{3.1.8}$$

The outcome is

$$\begin{aligned}\langle x,t|x',t_0\rangle &= \frac{1}{2\pi\hbar}\sqrt{\frac{\pi}{\frac{\mathrm{i}}{\hbar}\frac{t-t_0}{2M}}}\,\mathrm{e}^{\frac{\left(\frac{\mathrm{i}}{\hbar}(x-x')\right)^2}{2\frac{\mathrm{i}}{\hbar}\frac{t-t_0}{2M}}} \\ &= \sqrt{\frac{M}{\mathrm{i}2\pi\hbar(t-t_0)}}\,\mathrm{e}^{\frac{\mathrm{i}}{\hbar}\frac{M}{2}\frac{(x-x')^2}{t-t_0}}\,.\end{aligned} \tag{3.1.9}$$

It is worth verifying that this is indeed a solution of the Schrödinger differential equation

$$\mathrm{i}\hbar\frac{\partial}{\partial t}\langle x,t|x',t_0\rangle = -\frac{\hbar^2}{2M}\frac{\partial^2}{\partial x^2}\langle x,t|x',t_0\rangle\,, \tag{3.1.10}$$

which check is easily performed, and that it obeys the initial condition (2.2.3),

$$\langle x,t|x',t_0\rangle \to \delta(x-x') \quad \text{as} \quad t\to t_0\,. \tag{3.1.11}$$

In this limit, $t-t_0$ becomes arbitrarily small, so that $e^{\frac{i}{\hbar}\frac{M}{2}\frac{(x-x')^2}{t-t_0}}$ goes through many oscillations over small ranges of x', thus effectively washing out any contribution to

$$\int dx'\, e^{\frac{i}{\hbar}\frac{M}{2}\frac{(x-x')^2}{t-t_0}}\, f(x')$$

from $x'\neq x$, and the only actual contribution to the integral originates in the vicinity of $x'=x$ where the argument of the exponential function is extremal. This contribution is multiplied by the prefactor $\propto (t-t_0)^{-1/2}$ which is exceedingly large in the limit $t\to t_0$. Thus, indeed, $\langle x,t|x',t_0\rangle$ has the characteristic features of $\delta(x-x')$ as $t\to t_0$. What remains to be checked is the correct normalization; see

$$\int dx' \sqrt{\frac{M}{i2\pi\hbar(t-t_0)}}\, e^{\frac{i}{\hbar}\frac{M}{2}\frac{(x-x')^2}{t-t_0}} = \sqrt{\frac{M}{i2\pi\hbar(t-t_0)}}\sqrt{\frac{\pi}{\frac{1}{i\hbar}\frac{M}{2}\frac{1}{t-t_0}}} = 1\,, \tag{3.1.12}$$

it is just another gaussian integral.

Let us look at the dependence of the px time transformation function $\langle p,t|x,t_0\rangle$ on its labels p,t and x,t_0. We have the t derivative in (3.1.4),

$$i\hbar\frac{\partial}{\partial t}\langle p,t|x,t_0\rangle = \frac{p^2}{2M}\langle p,t|x,t_0\rangle\,, \tag{3.1.13}$$

and the immediate statements

$$\begin{aligned} i\hbar\frac{\partial}{\partial p}\langle p,t|x,t_0\rangle &= \langle p,t|X(t)|x,t_0\rangle\,,\\ i\hbar\frac{\partial}{\partial x}\langle p,t|x,t_0\rangle &= \langle p,t|P(t_0)|x,t_0\rangle\,,\\ \text{and}\quad -i\hbar\frac{\partial}{\partial t_0}\langle p,t|x,t_0\rangle &= \langle p,t|\underbrace{H(t_0)}_{=\frac{1}{2M}P(t_0)^2}|x,t_0\rangle \end{aligned} \tag{3.1.14}$$

about the other derivatives. We can turn all the operators on the right into numbers if we express them in terms of $P(t)$ and $X(t_0)$, because these operators will meet their eigenbras or eigenkets and can then be replaced by the eigenvalues. With the solutions of Heisenberg's equations of motion (3.1.2) at hand in (3.1.3), we achieve this quite easily, inasmuch as

$$\begin{aligned} P(t_0) &= P(t)\,, \\ X(t) &= X(t_0) + \frac{t-t_0}{M}P(t_0) \\ &= X(t_0) + \frac{t-t_0}{M}P(t)\,. \end{aligned} \tag{3.1.15}$$

Accordingly, we get

$$\begin{aligned} \mathrm{i}\hbar\frac{\partial}{\partial p}\langle p,t|x,t_0\rangle &= \left(x + \frac{t-t_0}{M}p\right)\langle p,t|x,t_0\rangle\,, \\ \mathrm{i}\hbar\frac{\partial}{\partial x}\langle p,t|x,t_0\rangle &= p\langle p,t|x,t_0\rangle\,, \\ -\mathrm{i}\hbar\frac{\partial}{\partial t_0}\langle p,t|x,t_0\rangle &= \frac{p^2}{2M}\langle p,t|x,t_0\rangle\,, \end{aligned} \tag{3.1.16}$$

and all are easily checked against the explicit expression for $\langle p,t|x,t_0\rangle$ in (3.1.5). Note in particular that the last equation amounts to

$$\frac{\partial}{\partial t}\langle p,t|x,t_0\rangle = -\frac{\partial}{\partial t_0}\langle p,t|x,t_0\rangle\,. \tag{3.1.17}$$

3-1 Can you think of a general reason why this should be so?

As a simple, yet important application of the xx time transformation function, we use it to find $\psi(x,t)$ if $\psi(x,t_0)$ is the gaussian wave function of the minimum-uncertainty state, that is: $\delta X\delta P = \frac{1}{2}\hbar$. In accordance with (4.8.10) in *Basic Matters*, we thus have

$$\psi(x,t_0) = \frac{(2\pi)^{-1/4}}{\sqrt{\delta X}}\,\mathrm{e}^{-\left(\frac{x}{2\delta X}\right)^2} \tag{3.1.18}$$

where δX is the spread in position, $\delta X = \sqrt{\langle X^2\rangle - \langle X\rangle^2}$, at time t_0; the momentum spread is $\delta P = \frac{\hbar}{2}/\delta X$, of course. So,

$$\begin{aligned} \psi(x,t) &= \int \mathrm{d}x'\,\langle x,t|x',t_0\rangle\,\psi(x',t_0) \qquad (3.1.19) \\ &= \int \mathrm{d}x'\sqrt{\frac{M}{\mathrm{i}2\pi\hbar(t-t_0)}}\,\mathrm{e}^{\frac{\mathrm{i}}{\hbar}\frac{M}{2}\frac{(x-x')^2}{t-t_0}}\,\frac{(2\pi)^{-1/4}}{\sqrt{\delta X}}\,\mathrm{e}^{-\left(\frac{x}{2\delta X}\right)^2}. \end{aligned}$$

This is another gaussian integral, as we emphasize by isolating the various factors,

$$\begin{aligned}\psi(x,t) &= \frac{(2\pi)^{-1/4}}{\sqrt{\delta X}}\sqrt{\frac{M}{\mathrm{i}2\pi\hbar(t-t_0)}}\,\mathrm{e}^{\frac{\mathrm{i}}{\hbar}\frac{M}{2}\frac{x^2}{t-t_0}}\int \mathrm{d}x'\,\mathrm{e}^{-\alpha x'^2+\beta x'} \\ &= \frac{(2\pi)^{-1/4}}{\sqrt{\delta X}}\sqrt{\frac{M}{\mathrm{i}2\pi\hbar(t-t_0)}\frac{\pi}{\alpha}}\,\mathrm{e}^{\frac{\mathrm{i}}{\hbar}\frac{M}{2}\frac{x^2}{t-t_0}}\,\mathrm{e}^{\frac{\beta^2}{4\alpha}} \end{aligned} \tag{3.1.20}$$

with

$$\begin{aligned}\alpha &= \frac{1}{\mathrm{i}\hbar}\frac{M}{2}\frac{1}{t-t_0}+\frac{1}{(2\delta X)^2} \\ &= \frac{1}{\mathrm{i}\hbar}\frac{M}{2}\frac{1}{t-t_0}\left(1+\mathrm{i}\frac{\hbar/2}{(\delta X)^2}\frac{t-t_0}{M}\right) \\ &= \frac{1}{\mathrm{i}\hbar}\frac{M}{2}\frac{1}{t-t_0}\left(1+\mathrm{i}\frac{t-t_0}{M}\frac{\delta P}{\delta X}\right) \\ &\equiv \frac{1}{\mathrm{i}\hbar}\frac{M}{2}\frac{1}{t-t_0}\frac{1}{\epsilon(t)} \end{aligned} \tag{3.1.21}$$

and $\beta = \dfrac{1}{\mathrm{i}\hbar}\dfrac{Mx}{t-t_0}$, where

$$\epsilon(t) = \left(1+\mathrm{i}\frac{t-t_0}{M}\frac{\delta P}{\delta X}\right)^{-1} \tag{3.1.22}$$

is a convenient abbreviation. We meet it in the prefactor

$$\sqrt{\frac{M}{\mathrm{i}2\pi\hbar(t-t_0)}\frac{\pi}{\alpha}} = \sqrt{\epsilon(t)}\,, \tag{3.1.23}$$

and in the argument of the exponential function,

$$\begin{aligned}\frac{\mathrm{i}}{\hbar}\frac{M}{2}\frac{x^2}{t-t_0}+\frac{\beta^2}{4\alpha} &= \frac{\mathrm{i}}{\hbar}\frac{M}{2}\frac{x^2}{t-t_0}+\frac{\mathrm{i}\hbar(t-t_0)}{2M}\epsilon(t)\left(\frac{1}{\mathrm{i}\hbar}\frac{Mx}{t-t_0}\right)^2 \\ &= \frac{\mathrm{i}}{\hbar}\frac{M}{2}\frac{x^2}{t-t_0}\underbrace{\Bigl(1-\epsilon(t)\Bigr)}_{=\mathrm{i}\frac{t-t_0}{M}\frac{\delta P}{\delta X}\epsilon(t)} \\ &= -\frac{\epsilon(t)}{2\hbar}\frac{\delta P}{\delta X}x^2 = -\epsilon(t)\left(\frac{x}{2\delta X}\right)^2, \end{aligned} \tag{3.1.24}$$

where the last step exploits $\delta P = \frac{\hbar}{2}/\delta X$.

In summary we have the compact result

$$\psi(x,t) = \frac{(2\pi)^{-1/4}}{\sqrt{\delta X/\epsilon(t)}}\, \mathrm{e}^{-\epsilon(t)\left(\frac{x}{2\delta X}\right)^2} \tag{3.1.25}$$

for the time-dependent wave function. We square it to obtain the time-dependent probability density

$$|\psi(x,t)|^2 = \frac{(2\pi)^{-1/2}}{\delta X(t)}\, \mathrm{e}^{-\frac{1}{2}\left(\frac{x}{\delta X(t)}\right)^2} \tag{3.1.26}$$

where

$$\begin{aligned} \delta X(t) = \frac{\delta X}{|\epsilon(t)|} &= \delta X \sqrt{1 + \left(\frac{t-t_0}{M}\frac{\delta P}{\delta X}\right)^2} \\ &= \sqrt{(\delta X)^2 + \left(\frac{t-t_0}{M}\delta P\right)^2}\,. \end{aligned} \tag{3.1.27}$$

The relations

$$\mathrm{Re}(\epsilon(t)) = |\epsilon(t)|^2 = \frac{1}{1 + \left(\frac{t-t_0}{M}\frac{\delta P}{\delta X}\right)^2} \tag{3.1.28}$$

help to get this result.

3.1.1 *Time-dependent spreads*

The time-dependent position spread

$$\begin{aligned} \delta X(t) &= \sqrt{(\delta X)^2 + \left(\frac{t-t_0}{M}\delta P\right)^2} \\ &\cong \frac{t-t_0}{M}\delta P \quad \text{for} \quad t - t_0 \gg \frac{M\delta X}{\delta P} \end{aligned} \tag{3.1.29}$$

illustrates the so-called "spreading of the wave function", a phenomenon that always occurs for free-moving quantum objects. It is just a manifestation of the familiar fact that an initial momentum spread turns into a rather large position spread after a while, even if the initial position spread is small. The time scale for this spreading is set by the ratio $M\delta X/\delta P$, it happens the faster, the larger is the initial momentum spread δP, and wave functions of less massive objects spread more rapidly than those of more massive objects.

The result about $\delta X(t)$ is more generally true than the specific example of an initial gaussian wave function might suggest. To make the general point, we return to the solutions of Heisenberg's equations of motion in (3.1.3) and note that

$$\begin{aligned} P(t)^2 &= P(t_0)^2 \\ X(t)^2 &= X(t_0)^2 + \frac{t-t_0}{M}\Big(X(t_0)P(t_0) + P(t_0)X(t_0)\Big) \\ &\quad + \left(\frac{t-t_0}{M}P(t_0)\right)^2 , \end{aligned} \tag{3.1.30}$$

so that we have the expectation values

$$\begin{aligned} \langle P(t)\rangle &= \langle P(t_0)\rangle , \\ \langle X(t)\rangle &= \langle X(t_0)\rangle + \frac{t-t_0}{M}\langle P(t_0)\rangle \end{aligned} \tag{3.1.31}$$

for the operators, and

$$\begin{aligned} \left\langle P(t)^2\right\rangle &= \left\langle P(t_0)^2\right\rangle , \\ \left\langle X(t)^2\right\rangle &= \left\langle X(t_0)^2\right\rangle + \left(\frac{t-t_0}{M}\right)^2 \left\langle P(t_0)^2\right\rangle \\ &\quad + \frac{t-t_0}{M}\Big\langle\Big(X(t_0)P(t_0) + P(t_0)X(t_0)\Big)\Big\rangle \end{aligned} \tag{3.1.32}$$

for their squares. The time-dependent spreads of X and P are therefore related to those at the initial time by

$$\begin{aligned} \delta X(t)^2 &= \left\langle X(t)^2\right\rangle - \langle X(t)\rangle^2 \\ &= \delta X(t_0)^2 + \left(\frac{t-t_0}{M}\delta P(t_0)\right)^2 \\ &\quad + 2\frac{t-t_0}{M}\left(\left\langle \frac{1}{2}[X(t_0)P(t_0) + P(t_0)X(t_0)]\right\rangle \right. \\ &\qquad \left. - \langle X(t_0)\rangle\langle P(t_0)\rangle\right) , \\ \delta P(t)^2 &= \delta P(t_0)^2 , \end{aligned} \tag{3.1.33}$$

so there is no change in the momentum spread, which is as anticipated because no forces are acting that would transfer momentum. By contrast, the position spread does change in time, and at very late times it is given by

$$\delta X(t) \cong \frac{t-t_0}{M}\delta P(t_0) \quad \text{(at late times)} \tag{3.1.34}$$

confirming quite generally what we found in (3.1.29) for the particular situation of an initial minimum-uncertainty wave function. At intermediate times, the correlation term

$$2\frac{t-t_0}{M}\left(\left\langle\frac{1}{2}\left[X(t_0)P(t_0)+P(t_0)X(t_0)\right]\right\rangle - \langle X(t_0)\rangle\langle P(t_0)\rangle\right)$$
$$\equiv 2\frac{t-t_0}{M}C(t_0) \tag{3.1.35}$$

may be noticeable. The value of the *position-momentum correlation* $C(t_0)$ can be positive, or negative, or vanishing altogether.

3-2 Show that the third possibility is the case, that is: $C(t_0) = 0$, for the minimum-uncertainty wave function in (3.1.18). Why do you expect this?

3-3 Define the time-dependent position-momentum correlation $C(t)$ in accordance with

$$C(t) = \left\langle\frac{1}{2}\left[X(t)P(t)+P(t)X(t)\right]\right\rangle - \langle X(t)\rangle\langle P(t)\rangle$$

and express it in terms of expectation values at the initial time t_0.

3.1.2 *Uncertainty ellipse*

For a visualization of the "spreading of the wave function", we use a graphical representation in the x,p-plane, the so-called *phase space*. Since the metrical dimensions of position x and momentum p are different, we choose an arbitrary length scale a and the corresponding momentum scale $\hbar/a$ and deal with the dimensionless operators X/a for position and $aP/\hbar$ for momentum.

The pair

$$\Big(\langle X/a\rangle, \langle aP/\hbar\rangle\Big) = \Big(\langle X\rangle/a, a\langle P\rangle/\hbar\Big) \tag{3.1.36}$$

of their expectation values identifies the central point in the phase space for the given state of the system, marked by the cross + in this plot:

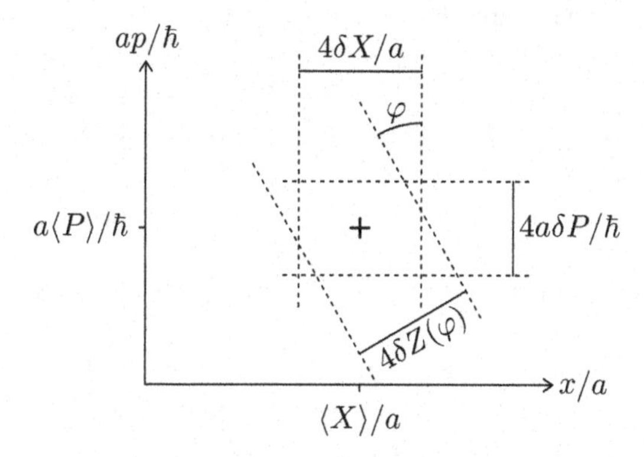

The vertical and horizontal dashed lines define the strips of width $4\delta X/a$ and $4a\delta P/\hbar$, respectively, that indicate the spreads in position and momentum, taking two standard deviations to each side of the mean value, so to say.

The horizontal and vertical directions are but two of many in phase space. Let us therefore consider the one-parameter family of operators $Z(\varphi)$ that we obtain by rotating the x axis toward the p axis,

$$Z(\varphi) = \bigl(X/a\bigr)\cos\varphi + \bigl(aP/\hbar\bigr)\sin\varphi\,. \tag{3.1.37}$$

The rotation angle φ specifies the chosen direction in phase space, with $\varphi = 0$ for the x axis and $\varphi = \frac{1}{2}\pi$ for the p axis.

3-4 Show that

$$\bigl[Z(\varphi_1), Z(\varphi_2)\bigr] = \mathrm{i}\sin(\varphi_2 - \varphi_1)$$

and conclude that

$$\delta Z(\varphi_1)\,\delta Z(\varphi_2) \geq \frac{1}{2}|\sin(\varphi_2 - \varphi_1)|$$

holds for the product of the spreads of $Z(\varphi)$ for any two rotation angles φ_1 and φ_2. Then determine the minimal value of this product for two perpendicular directions ($\varphi_2 = \varphi_1 + \frac{1}{2}\pi$, that is) to establish

$$\delta X\,\delta P \geq \frac{\hbar}{2}\sqrt{1 + (2C/\hbar)^2}\,.$$

This generalizes, and is more stringent than, Werner Heisenberg's famous uncertainty relation $\delta X\,\delta P \geq \dfrac{\hbar}{2}$.

3-5 For each value of φ, the eigenvalues of $Z(\varphi)$ are all real numbers z, with the respective eigenkets denoted by $|z,\varphi\rangle$,

$$Z(\varphi)|z,\varphi\rangle = |z,\varphi\rangle z\,.$$

Their orthonormality and completeness relations are

$$\langle z,\varphi|z',\varphi\rangle = \delta(z-z')\,, \qquad \int \mathrm{d}z\,|z,\varphi\rangle\langle z,\varphi| = 1$$

for each φ. Determine the position wave functions $\langle x|z,\varphi\rangle$.

For each value of φ, we also draw the two lines that define the strip of width $4\delta Z(\varphi)$ centered at the cross $+$ in the plot, whereby

$$\begin{aligned}\delta Z(\varphi) &= \sqrt{\left\langle Z(\varphi)^2\right\rangle - \langle Z(\varphi)\rangle^2} \\ &= \Big[(\delta X/a)^2(\cos\varphi)^2 + (a\delta P/\hbar)^2(\sin\varphi)^2 \\ &\qquad + (C/\hbar)\sin(2\varphi)\Big]^{\frac{1}{2}}\end{aligned} \tag{3.1.38}$$

involves the position spread δX, the momentum spread δP, and the position-momentum correlation C. Together, all these strips identify an area of the shape of an ellipse centered at the location (3.1.36) of the cross in the plot on page 61:

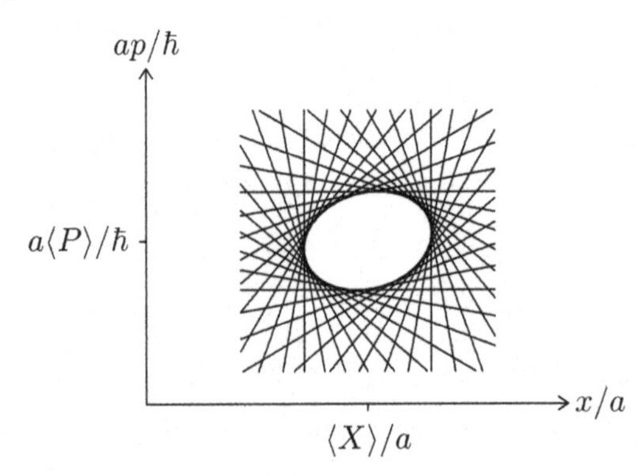

Point $(x/a, ap/\hbar)$ is on this *uncertainty ellipse* if

$$\left(\frac{x-\langle X\rangle}{2\delta X}\right)^2 + \left(\frac{p-\langle P\rangle}{2\delta P}\right)^2 - \frac{2C}{\delta X\,\delta P}\frac{x-\langle X\rangle}{2\delta X}\frac{p-\langle P\rangle}{2\delta P} = 1 - \left(\frac{C}{\delta X\,\delta P}\right)^2. \tag{3.1.39}$$

It follows that the area of the uncertainty ellipse, in units of $\hbar$, is

$$\frac{4\pi}{\hbar}\sqrt{\left(\delta X\,\delta P\right)^2 - C^2}\,. \tag{3.1.40}$$

We multiply by $a \times (\hbar/a) = \hbar$ to reinstall the metrical dimensions for X and P and get

$$\text{area} = 4\pi\sqrt{\left(\delta X\,\delta P\right)^2 - C^2}\,. \tag{3.1.41}$$

3-6 Show that this area is never less than $2\pi\hbar$.

In the course of time, the center of the uncertainty ellipse moves with constant velocity parallel to the x axis, as stated in (3.1.31), while the shape of the ellipse changes in accordance with what (3.1.33) and the solution of Exercise 3-4 tell us about the major and minor axes and their orientation. The area of the ellipse, however, does not change in time.

3-7 Show that the area of the uncertainty ellipse is indeed constant in time, and sketch the ellipses at different times.

3.2 Constant force

As a second example we consider the motion under the influence of a constant force F, for which the Hamilton operator is

$$H = \frac{1}{2M}P^2 - FX\,. \tag{3.2.1}$$

Heisenberg's equations of motion,

$$\begin{aligned} \frac{\mathrm{d}}{\mathrm{d}t}X &= \frac{\partial H}{\partial P} = \frac{1}{M}P\,,\\ \frac{\mathrm{d}}{\mathrm{d}t}P &= -\frac{\partial H}{\partial X} = F \end{aligned} \tag{3.2.2}$$

confirm that a constant force F is acting. In fact, these equations look exactly like their classical counterparts. And so do their solutions

$$\begin{aligned} P(t) &= P(t_0) + F(t - t_0) = P(t_0) + FT\,,\\ X(t) &= X(t_0) + \frac{t - t_0}{M}P(t_0) + \frac{F}{2M}(t - t_0)^2\\ &= X(t_0) + \frac{T}{M}P(t_0) + \frac{FT^2}{2M}\,, \end{aligned} \tag{3.2.3}$$

where we abbreviate the elapsed time by T,

$$T \equiv t - t_0 \,. \tag{3.2.4}$$

For the time dependence of the expectation values we find

$$\begin{aligned} \bigl\langle P(t) \bigr\rangle &= \bigl\langle P(t_0) \bigr\rangle + FT \,, \\ \bigl\langle X(t) \bigr\rangle &= \bigl\langle X(t_0) \bigr\rangle + \frac{T}{M} \bigl\langle P(t_0) \bigr\rangle + \frac{FT^2}{2M} \,, \end{aligned} \tag{3.2.5}$$

again of classical appearance. Their squares

$$\begin{aligned} \bigl\langle P(t) \bigr\rangle^2 &= \bigl\langle P(t_0) \bigr\rangle^2 + 2FT \bigl\langle P(t_0) \bigr\rangle + (FT)^2 \,, \\ \bigl\langle X(t) \bigr\rangle^2 &= \bigl\langle X(t_0) \bigr\rangle^2 + \frac{2T}{M} \bigl\langle X(t_0) \bigr\rangle \bigl\langle P(t_0) \bigr\rangle \\ &\quad + \left(\frac{T}{M} \bigl\langle P(t_0) \bigr\rangle \right)^2 + \frac{FT^2}{M} \bigl\langle X(t_0) \bigr\rangle \\ &\quad + \frac{FT^3}{M^2} \bigl\langle P(t_0) \bigr\rangle + \left(\frac{FT^2}{2M} \right)^2 \end{aligned} \tag{3.2.6}$$

are combined with the expectation values of

$$\begin{aligned} P(t)^2 &= P(t_0)^2 + 2FT P(t_0) + (FT)^2 \,, \\ X(t)^2 &= X(t_0)^2 + \frac{T}{M} \Bigl(X(t_0)P(t_0) + P(t_0)X(t_0) \Bigr) \\ &\quad + \left(\frac{T}{M} P(t_0) \right)^2 + \frac{FT^2}{M} X(t_0) \\ &\quad + \frac{FT^3}{M^2} P(t_0) + \left(\frac{FT^2}{2M} \right)^2 \end{aligned} \tag{3.2.7}$$

to produce

$$\begin{aligned} \delta P(t)^2 &= \delta P(t_0)^2 \,, \\ \delta X(t)^2 &= \delta X(t_0)^2 + \frac{2T}{M} C(t_0) + \left(\frac{T}{M} \right)^2 \delta P(t_0)^2 \,, \end{aligned} \tag{3.2.8}$$

with the initial position-momentum correlation $C(t_0)$ of (3.1.35). We note that these equations show no sign of the force F, which is to say that the evolution of the spreads in X and P, as well as of the correlation between them, is not affected by the force. The force is noticeable in the mean momentum $\bigl\langle P(t) \bigr\rangle$ and the mean position $\bigl\langle X(t) \bigr\rangle$, but not in their spreads. As a consequence, the uncertainty ellipse of Section 3.1.2 changes shape exactly as it does for force-free motion, but its center follows the parabolic trajectory specified by the expectation values in (3.2.5).

Since the Hamilton operator of (3.2.1) is quadratic in P but only linear in X, the Schrödinger equation for $\langle p,t|$ will be simpler than that for $\langle x,t|$. So we turn to

$$\begin{aligned} \mathrm{i}\hbar\frac{\partial}{\partial t}\langle p,t| &= \langle p,t|\left(\frac{1}{2M}P(t)^2 - FX(t)\right) \\ &= \left(\frac{p^2}{2M} - \mathrm{i}\hbar F\frac{\partial}{\partial p}\right)\langle p,t|\,. \end{aligned} \tag{3.2.9}$$

The first step on the way to its solution is the introduction of an integrating factor on the right,

$$\mathrm{i}\hbar\frac{\partial}{\partial t}\langle p,t| = \mathrm{e}^{-\frac{\mathrm{i}}{\hbar}\frac{p^3}{6MF}}\left(-\mathrm{i}\hbar F\frac{\partial}{\partial p}\right)\mathrm{e}^{\frac{\mathrm{i}}{\hbar}\frac{p^3}{6MF}}\langle p,t|\,, \tag{3.2.10}$$

followed by bringing the first factor over to the left and removing the common $\mathrm{i}\hbar$ factor. At this stage we have

$$\frac{\partial}{\partial t}\left(\mathrm{e}^{\frac{\mathrm{i}}{\hbar}\frac{p^3}{6MF}}\langle p,t|\right) = -F\frac{\partial}{\partial p}\left(\mathrm{e}^{\frac{\mathrm{i}}{\hbar}\frac{p^3}{6MF}}\langle p,t|\right). \tag{3.2.11}$$

This partial differential equation is of the simple form

$$\frac{\partial}{\partial t}f(p,t) = -F\frac{\partial}{\partial p}f(p,t) \tag{3.2.12}$$

with the immediate solution

$$f(p,t) = g(p - Ft) \tag{3.2.13}$$

where $g(\ )$ is an arbitrary single-argument function. It is linked to the initial condition at $t = t_0$ by

$$f(p,t_0) = g(p - Ft_0) \tag{3.2.14}$$

so that

$$g(p) = f(p + Ft_0, t_0) \tag{3.2.15}$$

and

$$\begin{aligned} f(p,t) &= f\big((p - Ft) + Ft_0, t_0\big) \\ &= f(p - FT, t_0) \quad \text{with} \quad T = t - t_0\,. \end{aligned} \tag{3.2.16}$$

Accordingly, we get

$$\mathrm{e}^{\frac{\mathrm{i}}{\hbar}\frac{p^3}{6MF}}\big\langle p,t\big| = \mathrm{e}^{\frac{\mathrm{i}}{\hbar}\frac{(p-FT)^3}{6MF}}\big\langle p-FT,t_0\big| \,, \tag{3.2.17}$$

and then

$$\big\langle p,t\big| = \mathrm{e}^{\frac{\mathrm{i}}{\hbar}\frac{(p-FT)^3-p^3}{6MF}}\big\langle p-FT,t_0\big| \tag{3.2.18}$$

is the solution of the Schrödinger equation (3.2.9).

We match it with $\big|p',t_0\big\rangle$ to obtain the *pp* time transformation function

$$\begin{aligned}\big\langle p,t\big|p',t_0\big\rangle &= \mathrm{e}^{\frac{\mathrm{i}}{\hbar}\frac{(p-FT)^3-p^3}{6MF}}\,\delta(p-FT-p') \\ &= \mathrm{e}^{\frac{\mathrm{i}}{\hbar}\frac{p'^3-p^3}{6MF}}\,\delta(p-FT-p')\,.\end{aligned} \tag{3.2.19}$$

It implies that the momentum wave function $\psi(p,t)$ is related to that at t_0 by

$$\begin{aligned}\psi(p,t) &= \int \mathrm{d}p'\,\big\langle p,t\big|p',t_0\big\rangle\,\psi(p',t_0) \\ &= \mathrm{e}^{\frac{\mathrm{i}}{\hbar}\frac{(p-FT)^3-p^3}{6MF}}\,\psi(p-FT,t_0)\,,\end{aligned} \tag{3.2.20}$$

which we could also have obtained by just putting the state ket $\big|\;\big\rangle$ into (3.2.18), the equation for $\big\langle p,t\big|$. We note that the phase factor disappears when we ask for probabilities,

$$|\psi(p,t)|^2 = |\psi(p-FT,t_0)|^2\,. \tag{3.2.21}$$

This relation is as it should be: the distribution as a whole is shifted by FT, the momentum transferred by the force acting for time T, without however changing the shape of the distribution, as we know already that the momentum spread does not change in time, $\delta P(t) = \delta P(t_0)$. So the picture is like this:

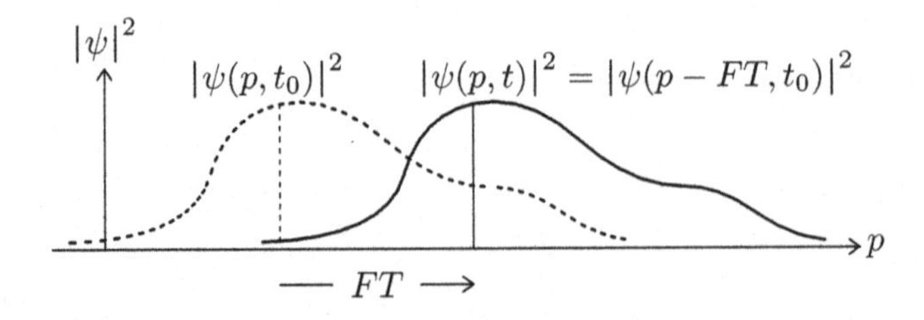

Next, we extract the px time transformation function

$$\begin{aligned}\langle p,t|x,t_0\rangle &= \mathrm{e}^{\frac{\mathrm{i}}{\hbar}\frac{(p-FT)^3-p^3}{6MF}}\langle p-FT,t_0|x,t_0\rangle\\ &= \frac{\mathrm{e}^{-\mathrm{i}(p-FT)x/\hbar}}{\sqrt{2\pi\hbar}}\,\mathrm{e}^{\frac{\mathrm{i}}{\hbar}\frac{(p-FT)^3-p^3}{6MF}}\\ &= \frac{\mathrm{e}^{-\mathrm{i}(p-FT)x/\hbar}}{\sqrt{2\pi\hbar}}\,\mathrm{e}^{-\frac{\mathrm{i}}{\hbar}\frac{T}{2M}\left(p-\frac{1}{2}FT\right)^2}\,\mathrm{e}^{-\frac{\mathrm{i}}{\hbar}\frac{F^2T^3}{24M}}\,. \qquad (3.2.22)\end{aligned}$$

3-8 Now find the xx time transformation function $\langle x,t|x',t_0\rangle$ by a suitable Fourier integration.

3-9 What is $\langle x,t|p,t_0\rangle$? (This is simple!) Use it to find $\psi(x,t)$ for

$$\psi(p,t_0) = \frac{(2\pi)^{-1/4}}{\sqrt{\delta P}}\,\mathrm{e}^{-\left(\frac{p}{2\delta P}\right)^2}\,. \qquad (3.2.23)$$

Do you get the expected result for $F \to 0$?

3.3 Time-dependent force

If the force is time dependent, $F = F(t)$, the Hamilton operator acquires a parametric time dependence,

$$H = \frac{1}{2M}P^2 - F(t)X\,, \qquad (3.3.1)$$

but otherwise we can proceed very similarly to the constant-force case. The Heisenberg equations of motion are now

$$\frac{\mathrm{d}}{\mathrm{d}t}P(t) = F(t)\,, \qquad (3.3.2)$$

solved by

$$P(t) = P(t_0) + \int_{t_0}^{t}\mathrm{d}t'\,F(t')\,, \qquad (3.3.3)$$

and

$$\frac{\mathrm{d}}{\mathrm{d}t}X(t) = \frac{1}{M}P(t)\,, \qquad (3.3.4)$$

solved by

$$
\begin{aligned}
X(t) &= X(t_0) + \int_{t_0}^{t} \mathrm{d}t''\, P(t'')/M \\
&= X(t_0) + \frac{t-t_0}{M}P(t_0) + \frac{1}{M}\int_{t_0}^{t} \mathrm{d}t'' \int_{t_0}^{t''} \mathrm{d}t'\, F(t')\,. \qquad (3.3.5)
\end{aligned}
$$

In this double integral, we have $t_0 < t' < t'' < t$, that is: both t' and t'' cover $t_0 \cdots t$ with $t' < t''$. Accordingly, we can interchange the order of integration if we pay due attention to the integration limits,

$$
\begin{aligned}
\int_{t_0}^{t} \mathrm{d}t'' \int_{t_0}^{t''} \mathrm{d}t'\, F(t') &= \int_{t_0}^{t} \mathrm{d}t' \int_{t'}^{t} \mathrm{d}t''\, F(t') \\
&= \int_{t_0}^{t} \mathrm{d}t'\, (t-t')F(t')\,, \qquad (3.3.6)
\end{aligned}
$$

and so arrive at

$$
X(t) = X(t_0) + \frac{t-t_0}{M}P(t_0) + \frac{1}{M}\int_{t_0}^{t} \mathrm{d}t'\, (t-t')F(t')\,. \qquad (3.3.7)
$$

Of course, one verifies easily that (3.3.4) is obeyed.

A compact way of presenting these equations is

$$
\begin{aligned}
P(t) &= P(t_0) + \Delta p(t,t_0)\,, \\
X(t) &= X(t_0) + \frac{t-t_0}{M}P(t_0) + \Delta x(t,t_0)\,, \qquad (3.3.8)
\end{aligned}
$$

with

$$
\Delta p(t,t_0) = \int_{t_0}^{t} \mathrm{d}t'\, F(t') \qquad (3.3.9)
$$

and

$$
\Delta x(t,t_0) = \frac{1}{M}\int_{t_0}^{t} \mathrm{d}t'\, (t-t')F(t')\,. \qquad (3.3.10)
$$

In (3.2.3) we have the constant-force expressions

$$
\begin{aligned}
\Delta p(t,t_0) &= FT\,, \\
\Delta x(t,t_0) &= \frac{FT^2}{2M} \quad \text{with} \quad T = t - t_0\,, \qquad (3.3.11)
\end{aligned}
$$

which are, of course, special cases of the more general expressions. These *purely numerical* terms do not enter the expressions of $\delta X(t)$ and $\delta P(t)$

in (3.2.8), and therefore we can conclude immediately that these relations remain true even when the force depends on time.

We illustrate another method for finding time transformation functions at the example of $\langle p,t|x,t_0\rangle$. We regard it as a function of all four labels and consider infinitesimal changes of each of them. Begin with p,

$$\mathrm{i}\hbar\frac{\partial}{\partial p}\langle p,t|x,t_0\rangle = \langle p,t|X(t)|x,t_0\rangle\,. \tag{3.3.12}$$

To proceed we use the solutions (3.3.8) of Heisenberg's equations of motion to express $X(t)$ in terms of $P(t)$ and $X(t_0)$,

$$\begin{aligned} X(t) &= X(t_0) + \frac{T}{M}P(t_0) + \Delta x(t,t_0) \\ &= X(t_0) + \frac{T}{M}\Big(P(t) - \Delta p(t,t_0)\Big) + \Delta x(t,t_0) \\ &= X(t_0) + \frac{T}{M}P(t) + \left(\Delta x - \frac{T}{M}\Delta p\right), \end{aligned} \tag{3.3.13}$$

and then arrive at

$$\mathrm{i}\hbar\frac{\partial}{\partial p}\langle p,t|x,t_0\rangle = \left(x + \frac{T}{M}p + \left(\Delta x - \frac{T}{M}\Delta p\right)\right)\langle p,t|x,t_0\rangle \tag{3.3.14}$$

or

$$\mathrm{i}\hbar\frac{\partial}{\partial p}\log\langle p,t|x,t_0\rangle = x + \frac{T}{M}p + \Delta x - \frac{T}{M}\Delta p\,. \tag{3.3.15}$$

Next, the dependence on x,

$$\mathrm{i}\hbar\frac{\partial}{\partial x}\langle p,t|x,t_0\rangle = \langle p,t|P(t_0)|x,t_0\rangle\,, \tag{3.3.16}$$

where $P(t_0) = P(t) - \Delta p(t,t_0)$ gives

$$\mathrm{i}\hbar\frac{\partial}{\partial x}\langle p,t|x,t_0\rangle = (p - \Delta p)\langle p,t|x,t_0\rangle \tag{3.3.17}$$

or

$$\mathrm{i}\hbar\frac{\partial}{\partial x}\log\langle p,t|x,t_0\rangle = p - \Delta p\,. \tag{3.3.18}$$

Now, the dependence on t,

$$\mathrm{i}\hbar\frac{\partial}{\partial t}\langle p,t|x,t_0\rangle = \langle p,t|H(t)|x,t_0\rangle \tag{3.3.19}$$

with the Hamilton operator at time t given by

$$\begin{aligned} H(t) &= \frac{1}{2M}P(t)^2 - F(t)X(t) \\ &= \frac{1}{2M}P(t)^2 - F(t)\Big(X(t_0) + \frac{T}{M}P(t) + \Delta x(t,t_0) - \frac{T}{M}\Delta p(t,t_0)\Big) \\ &\to \frac{p^2}{2M} - F(t)\Big(x + \frac{T}{M}p + \Delta x - \frac{T}{M}\Delta p\Big)\,, \end{aligned} \tag{3.3.20}$$

where the latter form is the numerical equivalent if bra $\langle p,t|$ and ket $|x,t_0\rangle$ are applied. So,

$$\mathrm{i}\hbar\frac{\partial}{\partial t}\log\langle p,t|x,t_0\rangle = \frac{p^2}{2M} - F(t)\Big(x + \frac{T}{M}p + \Delta x - \frac{T}{M}\Delta p\Big)\,. \tag{3.3.21}$$

Finally, the dependence on t_0,

$$-\mathrm{i}\hbar\frac{\partial}{\partial t_0}\langle p,t|x,t_0\rangle = \langle p,t|H(t_0)|x,t_0\rangle \tag{3.3.22}$$

with

$$\begin{aligned} H(t_0) &= \frac{1}{2M}P(t_0)^2 - F(t_0)X(t_0) \\ &= \frac{1}{2M}[P(t) - \Delta p(t,t_0)]^2 - F(t_0)X(t_0) \\ &\to \frac{1}{2M}(p-\Delta p)^2 - F(t_0)x\,, \end{aligned} \tag{3.3.23}$$

so that

$$\mathrm{i}\hbar\frac{\partial}{\partial t_0}\log\langle p,t|x,t_0\rangle = -\frac{(p-\Delta p)^2}{2M} + F(t_0)x\,. \tag{3.3.24}$$

These are four differential equations for *one* function of p, x, t, and t_0. It is therefore expedient to deal with all four equations at once, which we achieve by considering the response of $\langle p,t|x,t_0\rangle$ to simultaneous independent changes of all labels,

$$\delta\langle p,t|x,t_0\rangle = \Big(\delta x\frac{\partial}{\partial x} + \delta p\frac{\partial}{\partial p} + \delta t\frac{\partial}{\partial t} + \delta t_0\frac{\partial}{\partial t_0}\Big)\langle p,t|x,t_0\rangle\,, \tag{3.3.25}$$

or

$$\mathrm{i}\hbar\,\delta\log\langle p,t|x,t_0\rangle = \Big(\delta x\frac{\partial}{\partial x} + \cdots\Big)\mathrm{i}\hbar\log\langle p,t|x,t_0\rangle\,. \tag{3.3.26}$$

We are thus encountering

$$\begin{aligned}
\mathrm{i}\hbar\,\delta\log\langle p,t|x,t_0\rangle &= \delta p\Bigl(x+\frac{T}{M}p+\Delta x-\frac{T}{M}\Delta p\Bigr)\\
&\quad+\delta x(p-\Delta p)\\
&\quad+\delta t\Bigl(\frac{p^2}{2M}-F(t)\Bigl(x+\frac{T}{M}p+\Delta x-\frac{T}{M}\Delta p\Bigr)\Bigr)\\
&\quad+\delta t_0\Bigl(-\frac{(p-\Delta p)^2}{2M}+F(t_0)x\Bigr)\,. \qquad (3.3.27)
\end{aligned}$$

Since the left-hand side is a total variation, so must be the right-hand side. Indeed it is,

$$\mathrm{i}\hbar\,\delta\log\langle p,t|x,t_0\rangle = \delta\Bigl((p-\Delta p)\Bigl(x+\Delta x-\frac{T}{M}\Delta p\Bigr)+\frac{p^2T}{2M}-\hbar\Phi\Bigr) \qquad (3.3.28)$$

where

$$\Phi(t,t_0)=\frac{1}{2\hbar M}\int_{t_0}^{t}\mathrm{d}t'\int_{t_0}^{t}\mathrm{d}t''\,(t_>-t_0)F(t')F(t'') \qquad (3.3.29)$$

with

$$t_>=\mathrm{Max}\{t',t''\}=\begin{cases} t' & \text{if} \quad t'>t''\,,\\ t'' & \text{if} \quad t'<t''\,,\end{cases} \qquad (3.3.30)$$

is a force-dependent phase.

3-10 Establish first that

$$\begin{aligned}
\delta(\Delta p) &= \delta t F(t)-\delta t_0 F(t_0)\,,\\
\delta(\Delta x) &= \delta t\frac{1}{M}\Delta p-\delta t_0\frac{T}{M}F(t_0)\,,\\
\delta(\hbar\Phi) &= \delta t\frac{T}{M}\Delta p F(t)-\delta t_0\Bigl(\frac{(\Delta p)^2}{2M}-\Bigl(\Delta x-\frac{T}{M}\Delta p\Bigr)F(t_0)\Bigr)\,,
\end{aligned}$$

and verify then that the right-hand side of (3.3.28) has been identified correctly.

It now follows that

$$\langle p,t|x,t_0\rangle=\frac{1}{\sqrt{2\pi\hbar}}\exp\Bigl(-\frac{\mathrm{i}}{\hbar}(p-\Delta p)\Bigl(x+\Delta x-\frac{T}{M}\Delta p\Bigr)-\frac{\mathrm{i}}{\hbar}\frac{p^2T}{2M}+\mathrm{i}\Phi\Bigr) \qquad (3.3.31)$$

where the prefactor ensures the correct $t \to t_0$ limit,

$$\langle p, t | x, t_0 \rangle \to \frac{1}{\sqrt{2\pi\hbar}} e^{-ipx/\hbar} \quad \text{as } t \to t_0 . \tag{3.3.32}$$

In the limit of vanishing force, we should get the time transformation function for force-free motion, and we do indeed inasmuch as

$$\Delta x = 0 , \quad \Delta p = 0 , \quad \Phi = 0 \quad \text{for} \quad F \equiv 0 \tag{3.3.33}$$

and the result of (3.1.5) is recovered.

3-11 Find $\Phi(t, t_0)$ for a constant force $F(t) \equiv F$ and then verify that the result of (3.2.22) is correctly reproduced.

3-12 The state of the system is specified by the statistical operator at $t = t_0$,

$$\rho(X, P, t_0) = \rho_0(X, P)$$

What is $\rho(X, P, t)$ when a time-dependent force is acting?

3.4 Harmonic oscillator

After motion with no force, a constant force, and a time-dependent force, we now turn to the next more complicated dynamical system, that with a linear restoring force — a harmonic oscillator. Its Hamilton operator is

$$H = \frac{1}{2M} P^2 + \frac{1}{2} M\omega^2 X^2 , \tag{3.4.1}$$

and the Heisenberg equations of motion,

$$\begin{aligned} \frac{d}{dt} X &= \frac{1}{i\hbar}[X, H] = \frac{\partial H}{\partial P} = \frac{1}{M} P , \\ \frac{d}{dt} P &= \frac{1}{i\hbar}[P, H] = -\frac{\partial H}{\partial X} = -M\omega^2 X , \end{aligned} \tag{3.4.2}$$

are solved by

$$\begin{aligned} X(t) &= X(t_0) \cos\bigl(\omega(t - t_0)\bigr) + \frac{1}{M\omega} P(t_0) \sin\bigl(\omega(t - t_0)\bigr) , \\ P(t) &= P(t_0) \cos\bigl(\omega(t - t_0)\bigr) - M\omega X(t_0) \sin\bigl(\omega(t - t_0)\bigr) , \end{aligned} \tag{3.4.3}$$

as one can show by a variety of methods, at worst by verifying that they do obey the equations of motion.

The two time variables t and t_0 appear only as their difference $t - t_0$, and this is always multiplied by the frequency ω. It is therefore expedient to introduce the phase parameter

$$\phi = \omega(t - t_0) \tag{3.4.4}$$

as a convenient abbreviation. Then the solutions of the equations of motion read

$$\begin{aligned} X(t) &= X(t_0)\cos\phi + \frac{1}{M\omega}P(t_0)\sin\phi\,, \\ P(t) &= P(t_0)\cos\phi - M\omega X(t_0)\sin\phi\,. \end{aligned} \tag{3.4.5}$$

For later reference we note the commutator between position operators at different times,

$$\begin{aligned} \bigl[X(t), X(t_0)\bigr] &= \Bigl[X(t_0)\cos\phi + \frac{1}{M\omega}P(t_0)\sin\phi, X(t_0)\Bigr] \\ &= -\frac{\mathrm{i}\hbar}{M\omega}\sin\phi\,. \end{aligned} \tag{3.4.6}$$

3-13 What are the commutators $\bigl[X(t), P(t_0)\bigr]$ and $\bigl[P(t), X(t_0)\bigr]$ as well as $\bigl[P(t), P(t_0)\bigr]$?

We construct the xx time transformation function $\langle x, t | x', t_0\rangle$ with the aid of the method explained at (3.3.12)–(3.3.32), for which purpose we ask for the response to infinitesimal changes of all variables, that is x, x', t, and t_0. Begin with x and x',

$$\begin{aligned} \frac{\hbar}{\mathrm{i}}\frac{\partial}{\partial x}\langle x, t | x', t_0\rangle &= \langle x, t | P(t) | x', t_0\rangle\,, \\ \mathrm{i}\hbar\frac{\partial}{\partial x'}\langle x, t | x', t_0\rangle &= \langle x, t | P(t_0) | x', t_0\rangle\,, \end{aligned} \tag{3.4.7}$$

where we express $P(t)$ and $P(t_0)$, respectively, in terms of $X(t)$ and $X(t_0)$, which are the operators whose eigenbras and eigenkets are referred to in the time transformation function of present interest.

We have first

$$P(t_0) = \frac{M\omega}{\sin\phi}\Bigl(X(t) - X(t_0)\cos\phi\Bigr) \tag{3.4.8}$$

and then

$$
\begin{aligned}
P(t) &= \frac{M\omega}{\sin\phi}\Bigl(X(t) - X(t_0)\cos\phi\Bigr)\cos\phi - M\omega X(t_0)\sin\phi \\
&= \frac{M\omega}{\sin\phi}\Bigl(X(t)\cos\phi - X(t_0)\Bigr)\,.
\end{aligned} \tag{3.4.9}
$$

As a consequence,

$$
\begin{aligned}
\mathrm{i}\hbar\frac{\partial}{\partial x'}\langle x,t|x',t_0\rangle &= \langle x,t|\frac{M\omega}{\sin\phi}\Bigl(\underset{\displaystyle x}{\underset{\downarrow}{X(t)}} - \underset{\displaystyle x'}{\underset{\downarrow}{X(t_0)}}\cos\phi\Bigr)|x',t_0\rangle \\
&= \frac{M\omega}{\sin\phi}(x - x'\cos\phi)\langle x,t|x',t_0\rangle
\end{aligned} \tag{3.4.10}
$$

and likewise

$$
\frac{\hbar}{\mathrm{i}}\frac{\partial}{\partial x}\langle x,t|x',t_0\rangle = \frac{M\omega}{\sin\phi}(x\cos\phi - x')\langle x,t|x',t_0\rangle\,. \tag{3.4.11}
$$

The derivative with respect to t is given by the Schrödinger equation

$$
\mathrm{i}\hbar\frac{\partial}{\partial t}\langle x,t|x',t_0\rangle = \langle x,t|H(t)|x',t_0\rangle \tag{3.4.12}
$$

with the Hamilton operator

$$
\begin{aligned}
H(t) &= \frac{1}{2M}P(t)^2 + \frac{1}{2}M\omega^2X(t)^2 \\
&= \frac{1}{2M}\left(\frac{M\omega}{\sin\phi}\Bigl(X(t)\cos\phi - X(t_0)\Bigr)\right)^2 + \frac{1}{2}M\omega^2X(t)^2 \\
&= \frac{M\omega^2}{2(\sin\phi)^2}\Bigl(X(t)^2 + X(t_0)^2 - X(t)X(t_0)\cos\phi \\
&\qquad - X(t_0)X(t)\cos\phi\Bigr)\,.
\end{aligned} \tag{3.4.13}
$$

Except for the last term, all $X(t)$ are ready to be applied to $\langle x,t|$ and all $X(t_0)$ to $|x',t_0\rangle$, and this last term will be correctly ordered as soon as we take the commutator (3.4.6) into account,

$$
\begin{aligned}
X(t_0)X(t) &= X(t)X(t_0) - \bigl[X(t),X(t_0)\bigr] \\
&= X(t)X(t_0) + \frac{\mathrm{i}\hbar}{M\omega}\sin\phi\,.
\end{aligned} \tag{3.4.14}
$$

So,

$$H(t) = \frac{M\omega^2}{2(\sin\phi)^2}\Big(X(t)^2 + X(t_0)^2 - 2X(t)X(t_0)\cos\phi - \frac{i\hbar}{M\omega}\sin\phi\cos\phi\Big), \tag{3.4.15}$$

and in the context of (3.4.12), where $H(t)$ is sandwiched by $\langle x,t|$ and $|x',t_0\rangle$, it turns into a number,

$$i\hbar\frac{\partial}{\partial t}\langle x,t|x',t_0\rangle = \frac{M\omega^2}{2(\sin\phi)^2}\Big(x^2 + x'^2 - 2xx'\cos\phi - \frac{i\hbar}{M\omega}\sin\phi\cos\phi\Big)\langle x,t|x',t_0\rangle. \tag{3.4.16}$$

In the time derivative with respect to t_0,

$$-i\hbar\frac{\partial}{\partial t_0}\langle x,t|x',t_0\rangle = \langle x,t|H(t_0)|x',t_0\rangle, \tag{3.4.17}$$

we encounter $H(t_0)$, but that is the same as $H(t)$ because there is no parametric time dependence in the Hamilton operator.

3-14 Use the known solutions (3.4.5) of the equations of motion to verify explicitly that $H(t) = H(t_0)$, that is

$$\frac{1}{2M}P(t)^2 + \frac{1}{2}M\omega^2X(t)^2 = \frac{1}{2M}P(t_0)^2 + \frac{1}{2}M\omega^2X(t_0)^2.$$

In other words, $\langle x,t|x',t_0\rangle$ is not a function of t and t_0 individually, but depends only on the time difference $t - t_0 = \phi/\omega$, so that it will be convenient to switch from t and t_0 to ϕ,

$$\begin{aligned} i\hbar\frac{\partial}{\partial\phi}\langle x,t|x',t_0\rangle &= i\hbar\frac{1}{\omega}\frac{\partial}{\partial t}\langle x,t|x',t_0\rangle \\ &= \frac{M\omega}{2(\sin\phi)^2}\Big(x^2 + x'^2 - 2xx'\cos\phi - \frac{i\hbar}{M\omega}\sin\phi\cos\phi\Big)\langle x,t|x',t_0\rangle. \end{aligned} \tag{3.4.18}$$

We can now put the various pieces together and state the response of $\langle x,t|x',t_0\rangle$ to simultaneous independent infinitesimal variations of x, x', and ϕ,

$$\begin{aligned}\delta\log\langle x,t|x',t_0\rangle &= \left(\delta x\frac{\partial}{\partial x}+\delta x'\frac{\partial}{\partial x'}+\delta\phi\frac{\partial}{\partial\phi}\right)\log\langle x,t|x',t_0\rangle\\ &=\frac{\mathrm{i}}{\hbar}\left[\delta x\frac{M\omega}{\sin\phi}(x\cos\phi-x')-\delta x'\frac{M\omega}{\sin\phi}(x-x'\cos\phi)\right.\\ &\qquad\left.-\delta\phi\frac{M\omega}{2(\sin\phi)^2}\left(x^2+x'^2-2xx'\cos\phi-\frac{\mathrm{i}\hbar}{M\omega}\sin\phi\cos\phi\right)\right]\\ &=\frac{\mathrm{i}}{\hbar}\delta\left[\frac{M\omega}{2}(x^2+x'^2)\frac{\cos\phi}{\sin\phi}-M\omega xx'\frac{1}{\sin\phi}+\mathrm{i}\hbar\log\sqrt{\sin\phi}\right]\\ &=\delta\left[\frac{\mathrm{i}}{\hbar}\frac{M\omega}{2}(x^2+x'^2)\frac{\cos\phi}{\sin\phi}-\frac{\mathrm{i}}{\hbar}\frac{M\omega xx'}{\sin\phi}+\log\frac{1}{\sqrt{\sin\phi}}\right].\end{aligned}\tag{3.4.19}$$

It follows that $\log\langle x,t|x',t_0\rangle$ differs from $[\cdots]$ at most by an additive constant (meaning that it does not depend on x, x', or ϕ), so that there is a multiplicative constant in

$$\langle x,t|x',t_0\rangle=\frac{C}{\sqrt{\sin\phi}}\,\mathrm{e}^{\frac{\mathrm{i}}{\hbar}\frac{M\omega}{2}(x^2+x'^2)\frac{\cos\phi}{\sin\phi}-\frac{\mathrm{i}}{\hbar}\frac{M\omega xx'}{\sin\phi}}\,.\tag{3.4.20}$$

We can determine the value of C by considering the $\omega\to 0$ limit of force-free motion, when

$$\langle x,t|x',t_0\rangle\to\sqrt{\frac{M}{\mathrm{i}2\pi\hbar(t-t_0)}}\,\mathrm{e}^{\frac{\mathrm{i}}{\hbar}\frac{M}{2}\frac{(x-x')^2}{t-t_0}}\quad\text{as }\omega\to 0\,.\tag{3.4.21}$$

Since $\dfrac{\omega}{\sin\phi}\to\dfrac{1}{t-t_0}$ and $\omega\dfrac{\cos\phi}{\sin\phi}\to\dfrac{1}{t-t_0}$, the exponential factor in (3.4.20) has the correct limit all by itself, and the choice

$$\frac{C}{\sqrt{\sin\phi}}=\sqrt{\frac{M\omega}{\mathrm{i}2\pi\hbar\sin\phi}}\tag{3.4.22}$$

ensures the correct limit of the prefactor. In summary, then, we have established that

$$\langle x,t|x',t_0\rangle=\sqrt{\frac{M\omega}{\mathrm{i}2\pi\hbar\sin\phi}}\,\mathrm{e}^{\frac{\mathrm{i}}{\hbar}\frac{M\omega}{2\tan\phi}(x^2+x'^2)-\frac{\mathrm{i}}{\hbar}\frac{M\omega xx'}{\sin\phi}}\tag{3.4.23}$$

is the xx time transformation function for the harmonic oscillator.

3-15 Use the same method to derive the xp time transformation function $\langle x,t|p,t_0\rangle$ and verify that it is consistent with the xx time transformation function in (3.4.23).

In the limit $\omega \to 0$, the difference $x - x'$ is all that matters. We take this as a hint that it may be useful to express $\langle x, t|x', t_0\rangle$ in terms of $x - x'$ and $x + x'$. The identities

$$x^2 + x'^2 = \frac{1}{2}(x - x')^2 + \frac{1}{2}(x + x')^2 \tag{3.4.24}$$

and

$$xx' = -\frac{1}{4}(x - x')^2 + \frac{1}{4}(x + x')^2 \tag{3.4.25}$$

are then used to rewrite the argument of the exponential function,

$$\begin{aligned}
&\frac{M\omega}{2\tan\phi}(x^2 + x'^2) - \frac{M\omega x x'}{\sin\phi} \\
&= \frac{M\omega}{4\tan\phi}\left[(x - x')^2 + (x + x')^2\right] + \frac{M\omega}{4\sin\phi}\left[(x - x')^2 - (x + x')^2\right] \\
&= \frac{M\omega}{4\sin\phi}\left[(1 + \cos\phi)(x - x')^2 - (1 - \cos\phi)(x + x')^2\right] \\
&= \frac{1}{4}M\omega\left[(x - x')^2 \cot\frac{\phi}{2} - (x + x')^2 \tan\frac{\phi}{2}\right],
\end{aligned} \tag{3.4.26}$$

where the last step exploits the trigonometric identities

$$\frac{1 + \cos\phi}{\sin\phi} = \cot\frac{\phi}{2}, \qquad \frac{1 - \cos\phi}{\sin\phi} = \tan\frac{\phi}{2}. \tag{3.4.27}$$

Accordingly,

$$\begin{aligned}
\langle x, t|x', t_0\rangle = \sqrt{\frac{M\omega}{\mathrm{i}2\pi\hbar\sin\phi}}\; &\mathrm{e}^{\frac{\mathrm{i}}{\hbar}\frac{M}{2}(x - x')^2 \frac{\omega}{2}\cot\frac{\phi}{2}} \\
&\times \mathrm{e}^{-\frac{\mathrm{i}}{\hbar}\frac{M}{2}(x + x')^2 \frac{\omega}{2}\tan\frac{\phi}{2}}
\end{aligned} \tag{3.4.28}$$

where the $\omega \to 0$ limit is immediate inasmuch as ($T = t - t_0 = \phi/\omega$)

$$\begin{aligned}
&\frac{\omega}{2}\cot\frac{\phi}{2} = \frac{1}{T}\frac{\phi}{2}\cot\frac{\phi}{2} \to \frac{1}{T} \\
\text{and}\quad &\frac{\omega}{2}\tan\frac{\phi}{2} = \frac{1}{T}\frac{\phi}{2}\tan\frac{\phi}{2} \to 0 \quad \text{as } \phi \to 0\,.
\end{aligned} \tag{3.4.29}$$

We note that the short-time limit $T \to 0$ and the no-force limit $\omega \to 0$ are both tantamount to $\phi \to 0$, these limits are essentially the same. This is as it should be, because a force needs time to act to make itself felt, so that at very short times no effect of the force should be expected.

3-16 Find $\psi(x,t)$ for $\psi(x,t_0) = \dfrac{(2\pi)^{-1/4}}{\sqrt{\delta X}}\,\mathrm{e}^{-\left(\frac{x}{2\delta X}\right)^2}$.

3.4.1 *Ladder operators*

The Heisenberg equations of motion for the harmonic oscillator,

$$\frac{\mathrm{d}}{\mathrm{d}t}X = \frac{1}{M}P\,, \quad \frac{\mathrm{d}}{\mathrm{d}t}P = -M\omega^2 X\,, \tag{3.4.30}$$

can be combined into

$$\frac{\mathrm{d}}{\mathrm{d}t}(M\omega X + \mathrm{i}P) = -\mathrm{i}\omega(M\omega X + \mathrm{i}P) \tag{3.4.31}$$

or its adjoint

$$\frac{\mathrm{d}}{\mathrm{d}t}(M\omega X - \mathrm{i}P) = \mathrm{i}\omega(M\omega X - \mathrm{i}P)\,, \tag{3.4.32}$$

which are simple uncoupled equations of motion for the nonhermitian operators $M\omega X \pm \mathrm{i}P$ that are fully equivalent to the pair of coupled equations of motion for the hermitian operators X and P themselves. This simplification suggests strongly that also the Hamilton operator

$$H = \frac{1}{2M}P^2 + \frac{1}{2}M\omega^2 X^2 = \frac{1}{2M}\left[(M\omega X)^2 + P^2\right] \tag{3.4.33}$$

will look simpler if expressed in terms of these nonhermitian combinations of X and P. Indeed, if X and P were numbers, we would just have

$$(M\omega X)^2 + P^2 = (M\omega X + \mathrm{i}P)(M\omega X - \mathrm{i}P)\,, \tag{3.4.34}$$

but this is *not correct for operators*, because the right-hand side differs from the left-hand side by an additional term, namely

$$(M\omega X)(-\mathrm{i}P) + (\mathrm{i}P)(M\omega X) = -\mathrm{i}M\omega\underbrace{[X,P]}_{=\mathrm{i}\hbar} = M\hbar\omega\,. \tag{3.4.35}$$

Let us, therefore, be more systematic about this matter and take a look at the commutator

$$\begin{aligned}[][M\omega X + \mathrm{i}P, M\omega X - \mathrm{i}P] &= -\mathrm{i}M\omega\underbrace{[X,P]}_{=\mathrm{i}\hbar} + \mathrm{i}M\omega\underbrace{[P,X]}_{=-\mathrm{i}\hbar} \\ &= 2M\hbar\omega\,. \end{aligned} \tag{3.4.36}$$

It invites the definition of dimensionless nonhermitian operators

$$A = \frac{1}{\sqrt{2M\hbar\omega}}(M\omega X + \mathrm{i}P)\,,$$
$$A^\dagger = \frac{1}{\sqrt{2M\hbar\omega}}(M\omega X - \mathrm{i}P) \tag{3.4.37}$$

or, with

$$l = \sqrt{\frac{\hbar}{M\omega}} \quad \text{and} \quad \frac{\hbar}{l} = \sqrt{M\hbar\omega}\,, \tag{3.4.38}$$

which identify the length scale and the momentum scale of the harmonic oscillator,

$$A = \frac{1}{\sqrt{2}}\left(X/l + \mathrm{i}P\Big/\frac{\hbar}{l}\right),$$
$$A^\dagger = \frac{1}{\sqrt{2}}\left(X/l - \mathrm{i}P\Big/\frac{\hbar}{l}\right). \tag{3.4.39}$$

Solved for X and P, they appear as

$$X = l\,\frac{A + A^\dagger}{\sqrt{2}}\,, \qquad P = \frac{\hbar}{l}\,\frac{\mathrm{i}A^\dagger - \mathrm{i}A}{\sqrt{2}}\,, \tag{3.4.40}$$

which confirm that, in a manner of speaking, X is essentially the real part of A, and P the imaginary part.

Turning to the Hamilton operator, we note that

$$\begin{aligned} H &= \frac{1}{2M}\underbrace{\left(\frac{\hbar}{l}\right)^2\left(\frac{\mathrm{i}A^\dagger - \mathrm{i}A}{\sqrt{2}}\right)^2}_{=P^2} + \frac{1}{2}M\omega^2\underbrace{l^2\left(\frac{A + A^\dagger}{\sqrt{2}}\right)^2}_{=X^2} \\ &= \frac{1}{4}\hbar\omega\left[-(A^\dagger - A)^2 + (A + A^\dagger)^2\right] \end{aligned} \tag{3.4.41}$$

or

$$H = \frac{1}{2}\hbar\omega(A^\dagger A + AA^\dagger)\,. \tag{3.4.42}$$

This simplifies further as soon as we note that the fundamental commutator

$$\begin{aligned} [A, A^\dagger] &= \frac{1}{2}\left[\frac{X}{l} + \mathrm{i}\frac{P}{\hbar/l}, \frac{X}{l} - \mathrm{i}\frac{P}{\hbar/l}\right] \\ &= \frac{1}{2\hbar}\Big(\underbrace{[X, -\mathrm{i}P]}_{=\hbar} + \underbrace{[\mathrm{i}P, X]}_{=\hbar}\Big) = 1\,, \end{aligned} \tag{3.4.43}$$

that is simply

$$\left[A, A^\dagger\right] = 1\,, \tag{3.4.44}$$

enables us to express $AA^\dagger$ in terms of $A^\dagger A$,

$$AA^\dagger = A^\dagger A + 1\,, \tag{3.4.45}$$

so that

$$H = \hbar\omega A^\dagger A + \frac{1}{2}\hbar\omega\,. \tag{3.4.46}$$

Recalling the lesson of Section 3.3 in *Basic Matters*, we note that the additive constant $\frac{1}{2}\hbar\omega$ is of no real physical significance, inasmuch as it is completely irrelevant in the Heisenberg equation of motion (2.1.39),

$$\frac{\mathrm{d}}{\mathrm{d}t}F = \frac{\partial}{\partial t}F + \frac{1}{\mathrm{i}\hbar}[F, H]\,. \tag{3.4.47}$$

We can, therefore, simplify matters by dropping the $\frac{1}{2}\hbar\omega$,

$$H = \hbar\omega A^\dagger A + \frac{1}{2}\hbar\omega \to \hbar\omega A^\dagger A \tag{3.4.48}$$

or

$$H = \frac{1}{2M}P^2 + \frac{1}{2}M\omega^2 X^2 \to \frac{1}{2M}(M\omega X - \mathrm{i}P)(M\omega X + \mathrm{i}P)\,. \tag{3.4.49}$$

The new, slightly simplified, Hamilton operator is still a positive quantity — as is, of course, the original sum of squares — because all its expectation values are nonnegative,

$$\langle H\rangle = \langle\ |H|\ \rangle = \hbar\omega \underbrace{\langle\ |A^\dagger A|\ \rangle}_{\substack{\text{squared length}\\ \text{of ket } A|\ \rangle}} \geq 0\,. \tag{3.4.50}$$

It follows that

$$\left\langle \frac{P^2}{2M} + \frac{1}{2}M\omega^2 X^2 \right\rangle \geq \frac{1}{2}\hbar\omega \tag{3.4.51}$$

holds for the original Hamilton operator.

In these equations, the equal sign would apply to a ket that is an eigenket of A with eigenvalue 0,

$$A|0\rangle = |0\rangle 0 = 0\,. \tag{3.4.52}$$

The adjoint statement

$$\langle 0|A^\dagger = 0\langle 0| = 0 \tag{3.4.53}$$

identifies $\langle 0|$ as eigenbra of $A^\dagger$ with eigenvalue 0. Is there such an eigenket of A?

We answer this question by looking for its wave function $\langle x|0\rangle = \psi_0(x)$:

$$\begin{aligned} 0 = \langle x|A|0\rangle &= \langle x|\frac{1}{\sqrt{2}}\Bigl(\frac{X}{l} + \mathrm{i}\frac{l}{\hbar}P\Bigr)|0\rangle \\ &= \frac{1}{\sqrt{2}}\Bigl(\frac{x}{l} + l\frac{\partial}{\partial x}\Bigr)\langle x|0\rangle\,, \end{aligned} \tag{3.4.54}$$

stating that

$$l\frac{\partial}{\partial x}\psi_0(x) = -\frac{x}{l}\psi_0(x)\,. \tag{3.4.55}$$

This differential equation is solved by

$$\langle x|0\rangle = \psi_0(x) = \pi^{-1/4}l^{-1/2}\,\mathrm{e}^{-\frac{1}{2}(x/l)^2} \tag{3.4.56}$$

where we find the normalizing prefactor by a comparison with the known form of gaussian minimum-uncertainty wave functions,

$$\psi(x) = \frac{(2\pi)^{-1/4}}{\sqrt{\delta X}}\,\mathrm{e}^{-(\frac{1}{2}x/\delta X)^2}\,. \tag{3.4.57}$$

In passing, we establish that the position spread of $\psi_0(x)$ is

$$\delta X = \frac{1}{\sqrt{2}}l = \sqrt{\frac{\hbar}{2M\omega}} \tag{3.4.58}$$

and the momentum spread is then

$$\delta P = \frac{\hbar/2}{\delta X} = \frac{1}{\sqrt{2}}\frac{\hbar}{l} = \sqrt{\frac{1}{2}\hbar M\omega}\,. \tag{3.4.59}$$

Since

$$H|0\rangle = \hbar\omega A^\dagger \underbrace{A|0\rangle}_{=0} = 0\,, \tag{3.4.60}$$

we thus note that the ground state of the harmonic oscillator, the state with the lowest possible energy, is a minimum-uncertainty state with these spreads in position and momentum. It is in this sense that l sets the length scale for the harmonic oscillator, and $\hbar/l$ sets the momentum scale. We note further that $\hbar\omega$ must clearly be the natural unit of energy.

The latter observation is reinforced by noting that

$$\begin{aligned}(A^\dagger A)A^\dagger|0\rangle = A^\dagger(AA^\dagger)|0\rangle &= A^\dagger(A^\dagger A + 1)|0\rangle \\ &= A^\dagger|0\rangle\,, \end{aligned} \tag{3.4.61}$$

which states that $A^\dagger|0\rangle$ is eigenket of $A^\dagger A$ with eigenvalue 1. It is, therefore, eigenket of the Hamilton operator $H = \hbar\omega A^\dagger A$ with eigenvalue $\hbar\omega$.

We repeat this game,

$$\begin{aligned}(A^\dagger A)A^{\dagger 2}|0\rangle &= A^\dagger(AA^\dagger)A^\dagger|0\rangle \\ &= A^\dagger(\underbrace{A^\dagger A}_{\to 1}+1)A^\dagger|0\rangle = A^{\dagger 2}|0\rangle 2\,, \end{aligned} \tag{3.4.62}$$

stating that $A^{\dagger 2}|0\rangle$ is eigenket of $A^\dagger A$ with eigenvalue 2, and so forth by induction. We conclude that $A^{\dagger n}|0\rangle$ is eigenket of $A^\dagger A$ with eigenvalue n and verify the induction step:

$$\begin{aligned}(A^\dagger A)A^{\dagger n+1}|0\rangle &= A^\dagger(AA^\dagger)A^{\dagger n}|0\rangle \\ &= A^\dagger(\underbrace{A^\dagger A}_{\to n}+1)A^{\dagger n}|0\rangle \\ &= A^{\dagger n+1}|0\rangle(n+1)\,, \quad \text{indeed.} \end{aligned} \tag{3.4.63}$$

We want to have normalized eigenkets of $A^\dagger A$, and thus of $H = \hbar\omega A^\dagger A$, and so we need to establish the lengths of the kets $A^{\dagger n}|0\rangle$. We multiply by bra $\langle 0|A^n = \left(A^{\dagger n}|0\rangle\right)^\dagger$,

$$\begin{aligned}\langle 0|A^n A^{\dagger n}|0\rangle &= \langle 0|A^{n-1}\underbrace{(AA^\dagger)}_{=A^\dagger A + 1 \to (n-1)+1 = n} A^{\dagger n-1}|0\rangle \\ &= n\langle 0|A^{n-1}A^{\dagger n-1}|0\rangle\,, \end{aligned} \tag{3.4.64}$$

and by repeating this step $(n-1)$ more times,

$$\begin{aligned}\langle 0|A^n A^{\dagger n}|0\rangle &= n(n-1)(n-2)\cdots 2\cdot 1\underbrace{\langle 0|0\rangle}_{=1} \\ &= n!\,, \end{aligned} \tag{3.4.65}$$

so that

$$\big|n\big\rangle = \frac{A^{\dagger n}}{\sqrt{n!}}\big|0\big\rangle\,, \qquad n = 0, 1, 2, \ldots \tag{3.4.66}$$

and

$$\big\langle n\big| = \big\langle 0\big|\frac{A^n}{\sqrt{n!}}\,, \qquad n = 0, 1, 2, \ldots \tag{3.4.67}$$

are the normalized eigenkets and eigenbras of $A^\dagger A$, and thus of $H = \hbar\omega A^\dagger A$,

$$\begin{aligned} A^\dagger A\big|n\big\rangle &= \big|n\big\rangle n\,, \\ H\big|n\big\rangle &= \big|n\big\rangle\hbar\omega n\,. \end{aligned} \tag{3.4.68}$$

Part and parcel of the above construction are the relations

$$\begin{aligned} A^\dagger\big|n\big\rangle &= \big|n+1\big\rangle\sqrt{n+1}\,, \qquad & \big\langle n\big|A &= \sqrt{n+1}\big\langle n+1\big|\,, \\ A\big|n\big\rangle &= \big|n-1\big\rangle\sqrt{n}\,, \qquad & \big\langle n\big|A^\dagger &= \sqrt{n}\big\langle n-1\big|\,, \end{aligned} \tag{3.4.69}$$

which are the reasons why the operators A and $A^\dagger$ are called *ladder operators*: they take us up and down the ladder of $\big|n\big\rangle$ kets and $\big\langle n\big|$ bras from $n = 0$ to $n = 1, 2, \ldots$, rung by rung.

Could it be that $H = \hbar\omega A^\dagger A$ has some other eigenkets in addition to these just found? No, this is impossible as we see by *assuming* that there is some eigenket $\big|E\big\rangle, H\big|E\big\rangle = \big|E\big\rangle E$, with eigenvalue $E \neq n\hbar\omega$ for $n = 0, 1, 2, \ldots$. Then

$$\begin{aligned} HA\big|E\big\rangle &= \hbar\omega A^\dagger AA\big|E\big\rangle \\ &= \hbar\omega(AA^\dagger - 1)A\big|E\big\rangle \\ &= A\,\underbrace{\hbar\omega A^\dagger A\big|E\big\rangle}_{=\big|E\big\rangle E} - \hbar\omega A\big|E\big\rangle \\ &= A\big|E\big\rangle(E - \hbar\omega) \end{aligned} \tag{3.4.70}$$

states that $A\big|E\big\rangle$ is also eigenket of H with eigenvalue $E - \hbar\omega$. But then it follows that $A^2\big|E\big\rangle$ is eigenket with eigenvalue $E - 2\hbar\omega$ and so forth, eventually getting a negative eigenvalue. But that cannot be because all expectation values of H are nonnegative, as established in (3.4.50). In short, the assumption has an absurd consequence and so cannot be true.

3-17 Use

$$X^2 = \frac{l^2}{2}\left(A + A^\dagger\right)^2 = \frac{l^2}{2}\left(A^2 + AA^\dagger + A^\dagger A + A^{\dagger 2}\right)$$

to find $\langle n|X^2|n\rangle$ and then δX for $|n\rangle$. Similarly, find δP for $|n\rangle$. How large is their product $\delta X \delta P$?

3.4.2 *Coherent states*

We have found the eigenkets and bras of $A^\dagger A$, the so-called *Fock states* $|n\rangle$ and $\langle n|$, named after Vladimir A. Fock, but in fact it all began with a search for an eigenket of ladder operator A, with eigenvalue $A|0\rangle = 0$. How about other eigenvalues?

We try

$$A|a\rangle = |a\rangle a\,, \quad \text{where } a \text{ is some } \textit{complex} \text{ number,} \tag{3.4.71}$$

since the nonhermitian character of A suggests that its eigenvalues might be complex rather than real. Again, we look for the position wave function $\langle x|a\rangle$,

$$\begin{aligned}\langle x|A|a\rangle &= \langle x|\frac{1}{\sqrt{2}}\left(\frac{X}{l} + \mathrm{i}\frac{lP}{\hbar}\right)|a\rangle \\ &= \frac{1}{\sqrt{2}}\left(\frac{x}{l} + l\frac{\partial}{\partial x}\right)\langle x|a\rangle = \langle x|a\rangle a\,,\end{aligned} \tag{3.4.72}$$

or

$$l\frac{\partial}{\partial x}\langle x|a\rangle = \left(\frac{x}{l} + \sqrt{2}a\right)\langle x|a\rangle \tag{3.4.73}$$

so that

$$\langle x|a\rangle = \pi^{-1/4} l^{-1/2}\,\mathrm{e}^{-\frac{1}{2}\left(\frac{x}{l}\right)^2 + \sqrt{2}\frac{x}{l}a - \frac{1}{2}a^2}\,, \tag{3.4.74}$$

where the normalizing factor $\pi^{-1/4} l^{-1/2}\,\mathrm{e}^{-\frac{1}{2}a^2}$ turns into that of $\langle x|0\rangle$ in (3.4.56) when $a = 0$, and adopts a particular convention that we shall now explain.

For this purpose, let us note that the commutation relations

$$[X, P] = \mathrm{i}\hbar\,, \qquad \left[A, A^\dagger\right] = 1 \tag{3.4.75}$$

turn into each other under the replacements

$$X \to lA\,, \qquad P \to \mathrm{i}\frac{\hbar}{l}A^\dagger\,. \tag{3.4.76}$$

Indeed, (almost) any correct statement about X and P becomes a correct statement about A and $A^\dagger$ by these replacements, and vice versa. In particular, the differentiation rules

$$\begin{aligned}[X, f(X,P)] &= \mathrm{i}\hbar\frac{\partial}{\partial P}f(X,P)\,,\\ [f(X,P), P] &= \mathrm{i}\hbar\frac{\partial}{\partial X}f(X,P)\end{aligned} \tag{3.4.77}$$

have the analogs

$$\begin{aligned}\left[A, f(A^\dagger, A)\right] &= \frac{\partial}{\partial A^\dagger}f(A^\dagger, A)\,,\\ \left[f(A^\dagger, A), A^\dagger\right] &= \frac{\partial}{\partial A}f(A^\dagger, A)\,.\end{aligned} \tag{3.4.78}$$

3-18 Illustrate this for $f(A^\dagger, A) = (A^\dagger A)^2$.

And so we want, for example, that

$$P\big|x\big\rangle = \mathrm{i}\hbar\frac{\partial}{\partial x}\big|x\big\rangle \tag{3.4.79}$$

has the analog

$$A^\dagger\big|a\big\rangle = \frac{\partial}{\partial a}\big|a\big\rangle\,. \tag{3.4.80}$$

This is exactly ensured by the inclusion of the factor $\mathrm{e}^{-\frac{1}{2}a^2}$ in (3.4.74).

3-19 Verify this.

In addition, we wish that

$$\big\langle p\big|x\big\rangle = \frac{1}{\sqrt{2\pi\hbar}}\,\mathrm{e}^{-\mathrm{i}px/\hbar} = \frac{1}{\sqrt{2\pi\hbar}}\,\mathrm{e}^{\left(\frac{lp}{\mathrm{i}\hbar}\right)\left(\frac{x}{l}\right)} \tag{3.4.81}$$

has the analog

$$\big\langle a^*\big|a'\big\rangle = \mathrm{e}^{a^* a'} \tag{3.4.82}$$

where

$$\big\langle a^*\big| = \big|a\big\rangle^\dagger \quad \text{with} \quad a^* = \text{complex conjugate of } a \tag{3.4.83}$$

is the eigenbra of $A^\dagger$ with eigenvalue a^*,

$$\langle a^*|A^\dagger = a^*\langle a^*|\,, \tag{3.4.84}$$

that obtains from the eigenket $|a\rangle$ of A by taking the adjoint. There is no trace of the $\frac{1}{\sqrt{2\pi\hbar}}$ factor in (3.4.82), partly for simplicity, partly by convention, but mainly because we have earlier chosen to normalize $|0\rangle$ in accordance with $\langle 0|0\rangle = 1$.

To verify the statement about $\langle a^*|a'\rangle$ we first note that

$$\begin{aligned}\langle a^*|x\rangle &= \langle x|a\rangle^\dagger \\ &= \pi^{-1/4} l^{-1/2}\, \mathrm{e}^{-\frac{1}{2}a^{*2} + \sqrt{2}a^*\frac{x}{l} - \frac{1}{2}\left(\frac{x}{l}\right)^2}\end{aligned} \tag{3.4.85}$$

and then exploit the completeness of the x states in

$$\begin{aligned}\langle a^*|a'\rangle &= \int \mathrm{d}x\, \langle a^*|x\rangle\langle x|a'\rangle \\ &= \pi^{-1/2} l^{-1} \int \mathrm{d}x\; \mathrm{e}^{-\frac{1}{2}a^{*2} + \sqrt{2}(a^* + a')\frac{x}{l} - \left(\frac{x}{l}\right)^2 - \frac{1}{2}a'^2} \\ &= \pi^{-1/2} l^{-1}\, \mathrm{e}^{-\frac{1}{2}\left(a^{*2} + a'^2\right)} \sqrt{\pi}\, l\, \mathrm{e}^{\frac{(a^*+a')^2}{2}} \\ &= \mathrm{e}^{a^* a'}\,,\end{aligned} \tag{3.4.86}$$

where we recognize another gaussian integration.

As we have seen, there is a wealth of analogy between the hermitian pair X, P and the nonhermitian pair $A, A^\dagger$. How far does it go? Quite far, in fact, but it stops when we want to extend

$$|x\rangle \leftrightarrow |a\rangle\,, \qquad \langle p| \leftrightarrow \langle a^*| \qquad \text{(all right)} \tag{3.4.87}$$

to

$$\langle x| \leftrightarrow \langle a|\,, \qquad |p\rangle \leftrightarrow |a^*\rangle \qquad \text{(not all right)} \tag{3.4.88}$$

because ladder operator A has *no* eigenbras and ladder operator $A^\dagger$ has *no* eigenkets.

To make this point, assume an eigenbra $\langle a|$ of A, and look at the $\langle a|x\rangle$ wave function. It would have to obey the differential equation

$$\begin{aligned}a\langle a|x\rangle = \langle a|A|x\rangle &= \langle a|\frac{1}{\sqrt{2}}\left(\frac{X}{l} + \mathrm{i}\frac{lP}{\hbar}\right)|x\rangle \\ &= \frac{1}{\sqrt{2}}\left(\frac{x}{l} - l\frac{\partial}{\partial x}\right)\langle a|x\rangle\,,\end{aligned} \tag{3.4.89}$$

so that

$$\langle a|x\rangle \propto e^{\sqrt{2}a\frac{x}{l} + \frac{1}{2}(\frac{x}{l})^2} \tag{3.4.90}$$

and

$$\int dx\, \left|\langle a|x\rangle\right|^2 = \infty\,, \tag{3.4.91}$$

telling us that there is no such $\langle a|$.

With $|a\rangle$ being a function of the complex variable a, a function that can be differentiated,

$$A^\dagger |a\rangle = \frac{\partial}{\partial a}|a\rangle\,, \tag{3.4.92}$$

complex analysis tells us that $|a\rangle$ is an *entire function*, that is: a function that is analytic in the entire complex plane, which is of course a very restricted class of complex functions. The same applies to $\langle a^*|$, which is an entire function of a^*, and brakets such as

$$\langle a^*|a'\rangle = e^{a^*a'} \tag{3.4.93}$$

are, quite obviously, entire functions in both complex arguments.

3-20 Pretend that you do not know the eigenvalues of $A^\dagger A$. Proceed from $A^\dagger A|\nu\rangle = |\nu\rangle\nu$ (with temporarily unkown eigenvalue ν), establish that the entire function $\langle a^*|\nu\rangle$ obeys the differential equation

$$\left(a^*\frac{\partial}{\partial a^*} - \nu\right)\langle a^*|\nu\rangle = 0\,,$$

and infer that ν must be a nonnegative integer.

Accordingly, given an operator function $F(A^\dagger, A)$ of $A^\dagger$ and A, the mixed matrix element

$$\langle a^*|F(A^\dagger, A)|a'\rangle$$

is entire as a function of a' and also as a function of a^*. Upon multiplication with $e^{-a^*a'}$, we conclude that

$$\frac{\langle a^*|F(A^\dagger, A)|a'\rangle}{\langle a^*|a'\rangle} = f(a^*, a') \tag{3.4.94}$$

is analytical everywhere both in a^* and in a'. Since $f(A^\dagger; A)$ is clearly the $A^\dagger, A$-ordered version of F, it follows that all operators can be written in an

$A^\dagger, A$-ordered form and that this *normally ordered* form of F must be entire in $A^\dagger$ and in A. We have occasionally spoken of "reasonable functions" of X, P or other operators, and have now found a clear criterion for judging what is "reasonable": the normally ordered function is entire in $A^\dagger$ and A.

This possibility of normal ordering relies on the existence of eigenbras of $A^\dagger$ and eigenkets of A, and so there is no reason why one should be able to put any arbitrary operator into an $A, A^\dagger$-ordered form. Indeed, this *antinormal* ordering cannot be done for arbitrary operators, if they involve infinitely many As and infinitely many $A^\dagger$s.

Normal ordering can be a powerful tool, as is illustrated by the following exercise.

3-21 Show that $|0\rangle\langle 0| = \mathrm{e}^{-A^\dagger; A}$, and then extend this to

$$|n\rangle\langle n| = \frac{1}{n!} A^{\dagger n}\, \mathrm{e}^{-A^\dagger; A} A^n \,.$$

Use this to prove the completeness of the Fock states, that is

$$\sum_{n=0}^{\infty} |n\rangle\langle n| = 1 \,. \tag{3.4.95}$$

The kets $|a\rangle$ and bras $\langle a^*|$ are known as Roy J. Glauber's *coherent states*. They are complete as well — in fact, they are overcomplete, which is to say that there is more than one completeness relation, or equivalently, that subsets of these states are already complete.

3.4.3 *Completeness of the coherent states*

We recall the completeness relations for the x and p states and combine them such that we only have a $|p\rangle$ bra and a $\langle x|$ ket in the end, as they have $|a\rangle$ and $\langle a^*|$ as analogs:

$$\begin{aligned}
1 &= \int \mathrm{d}x\, |x\rangle\langle x| \int \mathrm{d}p\, |p\rangle\langle p| \\
&= \int \mathrm{d}x\, \mathrm{d}p\, |x\rangle \underbrace{\langle x|p\rangle}_{=\left(2\pi\hbar\langle p|x\rangle\right)^{-1}} \langle p| \\
&= \int \frac{\mathrm{d}x\, \mathrm{d}p}{2\pi\hbar} \frac{|x\rangle\langle p|}{\langle p|x\rangle}
\end{aligned} \tag{3.4.96}$$

and thus conjecture that the completeness relation of the coherent states $|a\rangle$ and $\langle a^*|$ should appear as

$$1 = \int \frac{\mathrm{d}x\,\mathrm{d}p}{2\pi\hbar} \frac{|a'\rangle\langle a^*|}{\langle a^*|a'\rangle}\bigg|_{\substack{\text{suitable}\\ \text{parameterization}}} . \tag{3.4.97}$$

The "suitable parameterization" is the injunction for relating the complex numbers a' and a^* to the real integration parameters x and p. There is, in fact, a plethora of possible parameterizations, they are all equally good on general grounds, but very often one is much more convenient than others for a particular application. The two most important, and most frequently used, parameterizations are

$$\begin{aligned} a' &= \frac{1}{\sqrt{2}}\left(\frac{x}{l} + \mathrm{i}\frac{lp}{\hbar}\right), \\ a^* &= \frac{1}{\sqrt{2}}\left(\frac{x}{l} - \mathrm{i}\frac{lp}{\hbar}\right) = a'^* , \end{aligned} \tag{3.4.98}$$

and

$$a' = \frac{x}{l} , \qquad a^* = -\mathrm{i}\frac{lp}{\hbar} . \tag{3.4.99}$$

The parameterization (3.4.98) is suggested by the basic relations (3.4.39) between A, $A^\dagger$ and X, P, whereas parameterization (3.4.99) is suggested by the useful analogy of (3.4.76). A third parameterization is the subject matter of Exercise 3-28 on page 92.

When using the parameterization (3.4.98), all eigenkets of A and all eigenbras of $A^\dagger$ contribute to the integral in (3.4.97). But if we employ the parameterization (3.4.99), only the eigenkets of A with real eigenvalues and the eigenbras of $A^\dagger$ with imaginary eigenvalues appear in the completeness relation. This implies that these subsets of kets and bras are already complete, and the whole set is overcomplete. Indeed, since $|a\rangle$ and $\langle a^*|$ are entire functions of their complex arguments, there are many subsets that are complete, among them subsets that are countable.

We leave the verification of parameterization (3.4.98) to Exercise 3-22 below, and demonstrate the parameterization (3.4.99) here by evaluating

$$\delta(x' - x'') = \langle x'|x''\rangle \stackrel{?}{=} \int \frac{\mathrm{d}x\,\mathrm{d}p}{2\pi\hbar} \frac{\langle x'|a'\rangle\langle a^*|x''\rangle}{\langle a^*|a'\rangle}\bigg|_{\substack{a' = x/l \\ a^* = -ilp/\hbar}} . \tag{3.4.100}$$

The wave functions in (3.4.74) and (3.4.82) give

$$\begin{aligned}\frac{\langle x'|a'\rangle\langle a^*|x''\rangle}{\langle a^*|a'\rangle} &= \frac{1}{\sqrt{\pi}l}\,\mathrm{e}^{-\frac{1}{2l^2}(x'^2+x''^2)}\,\mathrm{e}^{\sqrt{2}(\frac{x'}{l}a'+a^*\frac{x''}{l})}\\ &\quad\times\mathrm{e}^{-\frac{1}{2}a'^2-\frac{1}{2}a^{*2}}\,\mathrm{e}^{-a^*a'}\\ &= \frac{1}{\sqrt{\pi}l}\,\mathrm{e}^{-\frac{1}{2l^2}(x'^2+x''^2)}\,\mathrm{e}^{\sqrt{2}(\frac{x'x}{l^2}-\mathrm{i}x''p/\hbar)}\\ &\quad\times\mathrm{e}^{-\frac{1}{2}(\frac{x}{l})^2}\,\mathrm{e}^{\frac{1}{2}(\frac{lp}{\hbar})^2}\,\mathrm{e}^{\mathrm{i}xp/\hbar}\end{aligned}\qquad(3.4.101)$$

so that, upon evaluation of the gaussian x integral, we arrive at

$$\begin{aligned}\delta(x'-x'') &\overset{?}{=} \sqrt{2}\,\mathrm{e}^{-\frac{1}{2l^2}(x'^2-x''^2)}\underbrace{\int\frac{\mathrm{d}p}{2\pi\hbar}\,\mathrm{e}^{\mathrm{i}\sqrt{2}(x'-x'')p/\hbar}}_{=\delta(\sqrt{2}(x'-x''))}\\ &= \sqrt{2}\,\delta\big(\sqrt{2}(x'-x'')\big)\underbrace{\mathrm{e}^{\frac{1}{2l^2}(x'^2-x''^2)}}_{=1\quad\text{for}\quad x'=x''}\\ &= \delta(x'-x'')\,,\end{aligned}\qquad(3.4.102)$$

indeed.

3-22 Verify that (3.4.98) is a permissible parameterization in (3.4.97).

3-23 Show that

$$\mathrm{tr}\{F\} = \int\frac{\mathrm{d}x\,\mathrm{d}p}{2\pi\hbar}\,f\Big(-\mathrm{i}\frac{lp}{\hbar},\frac{x}{l}\Big)$$

where $F = f(A^\dagger, A)$ is the normally ordered form of operator F.

3.4.4 *Fock states and coherent states*

The Fock-state ket for $n = 0$ and the coherent-state ket for $a = 0$ are the same, $|0\rangle = |n=0\rangle = |a=0\rangle$. By combining (3.4.67) with (3.4.71) and (3.4.82) we can, therefore, infer immediately that the Fock states $|n\rangle$ and the coherent states $|a\rangle$ are related to each other by

$$\langle n|a\rangle = \langle 0|\frac{A^n}{\sqrt{n!}}|a\rangle = \langle 0|a\rangle\frac{a^n}{\sqrt{n!}} = \frac{a^n}{\sqrt{n!}}\qquad(3.4.103)$$

or

$$|a\rangle = \sum_{n=0}^{\infty} |n\rangle \frac{a^n}{\sqrt{n!}}\,, \tag{3.4.104}$$

which states that the coherent states constitute the generating function for the Fock states. This is demonstrated easily by verifying the basic relations $A|a\rangle = |a\rangle a$,

$$\begin{aligned} A|a\rangle &= \sum_{n=0}^{\infty} A|n\rangle \frac{a^n}{\sqrt{n!}} \\ &= \sum_{n=1}^{\infty} |n-1\rangle \sqrt{n} \frac{a^n}{\sqrt{n!}} = \sum_{n=0}^{\infty} |n\rangle \frac{a^{n+1}}{\sqrt{n!}} \\ &= |a\rangle a\,, \end{aligned} \tag{3.4.105}$$

and $A^\dagger |a\rangle = \dfrac{\partial}{\partial a}|a\rangle$,

$$\begin{aligned} A^\dagger |a\rangle &= \sum_{n=0}^{\infty} A^\dagger |n\rangle \frac{a^n}{\sqrt{n!}} \\ &= \sum_{n=0}^{\infty} |n+1\rangle \sqrt{n+1} \frac{a^n}{\sqrt{n!}} \\ &= \sum_{n=0}^{\infty} |n+1\rangle \frac{\partial}{\partial a} \frac{a^{n+1}}{\sqrt{(n+1)!}} \\ &= \frac{\partial}{\partial a} \sum_{n=0}^{\infty} |n\rangle \frac{a^n}{\sqrt{n!}} = \frac{\partial}{\partial a} |a\rangle\,, \end{aligned} \tag{3.4.106}$$

as well as $\langle a^* | a' \rangle = \mathrm{e}^{a^* a'}$,

$$\begin{aligned} \langle a^* | a' \rangle &= \sum_{n=0}^{\infty} \frac{a^{*n}}{\sqrt{n!}} \langle n | \sum_{m=0}^{\infty} |m\rangle \frac{a'^m}{\sqrt{m!}} \\ &= \sum_{n,m=0}^{\infty} \frac{a^{*n} a'^m}{\sqrt{n!\,m!}} \underbrace{\langle n | m \rangle}_{=\delta_{n,m}} \\ &= \sum_{n=0}^{\infty} \frac{a^{*n} a'^n}{n!} = \mathrm{e}^{a^* a'}\,, \end{aligned} \tag{3.4.107}$$

indeed.

3-24 Combine the $\langle x|a\rangle$ wave function of (3.4.74), with the completeness relation (3.4.95) of the Fock states and the generating function for the Hermite polynomials (Charles Hermite),

$$e^{2ty-t^2} = \sum_{n=0}^{\infty} \frac{t^n}{n!} \mathrm{H}_n(y),$$

to find the position wave functions $\langle x|n\rangle$ of the Fock states.

3-25 Provide yourself with the momentum wave function $\langle p|a\rangle$ of the coherent states and then follow the strategy of the preceding exercise to find the momentum wave functions $\langle p|n\rangle$ of the Fock states.

3-26 Use parameterization (3.4.98) to verify that the orthonormality of the Fock states, $\delta_{nm} = \langle n|m\rangle = \langle n|1|m\rangle$, is consistent with this completeness relation for the coherent states.

3-27 Do the same for parameterization (3.4.99).

3-28 Exploit the orthonormality of the Fock states once more to verify that

$$a' = s\,e^{i\phi}, \quad a^* = s\,e^{-i\phi} = a'^*, \quad dx\,dp = \hbar\,ds\,s\,d\phi$$

with $s \geq 0$ and $0 \leq \phi \leq 2\pi$ is another suitable parameterization for (3.4.97). How is it related to parameterization (3.4.98)?

3-29 Show that the normally ordered form of any function $F = f(A^\dagger A)$ of $A^\dagger A$ is given by

$$F = \sum_{n=0}^{\infty} \frac{f(n)}{n!} A^{\dagger n}\, e^{-A^\dagger; A} A^n .$$

Use this to find the normally ordered form of $(1-\lambda)^{A^\dagger A}$ with $\lambda > 0$.

3.4.5 *Time dependence*

The introduction of the nonhermitian operators of A and $A^\dagger$ was mainly motivated by the simplicity of their Heisenberg equations of motion,

$$\frac{d}{dt}A(t) = -i\omega A(t),$$
$$\frac{d}{dt}A^\dagger(t) = +i\omega A^\dagger(t), \qquad (3.4.108)$$

which are solved by

$$A(t) = \mathrm{e}^{-\mathrm{i}\omega(t - t_0)} A(t_0) ,$$
$$A^\dagger(t_0) = A^\dagger(t)\, \mathrm{e}^{-\mathrm{i}\omega(t - t_0)} , \tag{3.4.109}$$

respectively. In the latter we have chosen to express $A^\dagger$ at the earlier time t_0 by $A^\dagger$ at the later time t, rather than the other way around, because this is what we need in the variation

$$\delta\langle a^*, t|a', t_0\rangle = \left[\delta a^* \frac{\partial}{\partial a^*} + \delta a' \frac{\partial}{\partial a'} + \delta t \frac{\partial}{\partial t} + \delta t_0 \frac{\partial}{\partial t_0}\right] \langle a^*, t|a', t_0\rangle \tag{3.4.110}$$

of the time transformation function between coherent states. For, here

$$\begin{aligned}
\frac{\partial}{\partial a^*}\langle a^*, t|a', t_0\rangle &= \langle a^*, t|A(t)|a', t_0\rangle \\
&= \mathrm{e}^{-\mathrm{i}\omega(t - t_0)} a' \langle a^*, t|a', t_0\rangle , \\
\frac{\partial}{\partial a'}\langle a^*, t|a', t_0\rangle &= \langle a^*, t|A^\dagger(t_0)|a', t_0\rangle \\
&= a^*\, \mathrm{e}^{-\mathrm{i}\omega(t - t_0)} \langle a^*, t|a', t_0\rangle , \\
\frac{\partial}{\partial t}\langle a^*, t|a', t_0\rangle &= -\frac{\partial}{\partial t_0}\langle a^*, t|a', t_0\rangle \\
&= \frac{1}{\mathrm{i}\hbar}\langle a^*, t|H|a', t_0\rangle \\
&= -\mathrm{i}\omega a^*\, \mathrm{e}^{-\mathrm{i}\omega(t - t_0)} a' \langle a^*, t|a', t_0\rangle ,
\end{aligned} \tag{3.4.111}$$

where the relations (3.4.109) as well as their implication

$$H = \hbar\omega A^\dagger(t) A(t) = \hbar\omega A^\dagger(t)\, \mathrm{e}^{-\mathrm{i}\omega(t - t_0)} A(t_0) \tag{3.4.112}$$

are used. Upon putting the ingredients together, we have

$$\begin{aligned}
\delta\langle a^*, t|a', t_0\rangle = \Big[&\delta a^*\, \mathrm{e}^{-\mathrm{i}\omega(t - t_0)} a' + a^*\, \mathrm{e}^{-\mathrm{i}\omega(t - t_0)} \delta a' \\
&+ (\delta t - \delta t_0)(-\mathrm{i}\omega) a^*\, \mathrm{e}^{-\mathrm{i}\omega(t - t_0)} a'\Big] \langle a^*, t|a', t_0\rangle ,
\end{aligned} \tag{3.4.113}$$

or after dividing by $\langle a^*, t|a', t_0\rangle$,

$$\delta \log\langle a^*, t|a', t_0\rangle = \delta\Big(a^*\, \mathrm{e}^{-\mathrm{i}\omega(t - t_0)} a'\Big) . \tag{3.4.114}$$

It follows that

$$\langle a^*, t|a', t_0\rangle = \mathrm{e}^{a^*\, \mathrm{e}^{-\mathrm{i}\omega(t - t_0)} a'} , \tag{3.4.115}$$

where we have already identified correctly the multiplicative constant of integration such that the initial condition

$$\langle a^*, t | a', t_0 \rangle \to \mathrm{e}^{a^* a'} \quad \text{as } t \to t_0 \tag{3.4.116}$$

is obeyed.

We have thus found the time transformation function that relates eigenkets of $A(t_0)$ to eigenbras of $A^\dagger(t)$. In conjunction with either one of the completeness relations of the form (3.4.97), it enables us to find $\langle a^*, t | \ \rangle$ if $\langle a^*, t_0 | \ \rangle$ is given. In fact, we do not even need to exploit the completeness relations because

$$\begin{aligned} \langle a^*, t | a', t_0 \rangle &= \mathrm{e}^{a^* \mathrm{e}^{-\mathrm{i}\omega(t - t_0)} a'} \\ &= \mathrm{e}^{[a^* \mathrm{e}^{-\mathrm{i}\omega(t - t_0)}] a'} \\ &= \langle a^* \mathrm{e}^{-\mathrm{i}\omega(t - t_0)}, t_0 | a', t_0 \rangle \end{aligned} \tag{3.4.117}$$

so that, as a consequence of the completeness of the a' kets,

$$\langle a^*, t | = \langle a^* \mathrm{e}^{-\mathrm{i}\omega(t - t_0)}, t_0 | \,. \tag{3.4.118}$$

Therefore, if we know

$$\psi(a^*, t_0) = \langle a^*, t_0 | \ \rangle \,, \tag{3.4.119}$$

then $\psi(a^*, t) = \langle a^*, t | \ \rangle$ is immediately given by

$$\psi(a^*, t) = \psi(a^* \mathrm{e}^{-\mathrm{i}\omega(t - t_0)}, t_0) \,. \tag{3.4.120}$$

Tersely: The quantum number a^* acquires an additional complex phase factor $\mathrm{e}^{-\mathrm{i}\omega(t - t_0)}$, and that is *all*.

There are, of course, various ways of getting this result. For example, we could begin with first noting that the Schrödinger equation

$$\mathrm{i}\hbar \frac{\partial}{\partial t} \langle \ldots, t | = \langle \ldots, t | H \tag{3.4.121}$$

is formally solved by

$$\langle \ldots, t | = \langle \ldots, t_0 | \, \mathrm{e}^{-\frac{\mathrm{i}}{\hbar}(t - t_0) H} \tag{3.4.122}$$

provided that there is no parametric time dependence in the Hamilton operator, $\dfrac{\partial H}{\partial t} = 0$. Because then it follows that $\dfrac{\mathrm{d}H}{\mathrm{d}t} = 0$, so that there is one and the same Hamilton operator at all times. The eigenbras of H itself,

$$\langle E | H = E \langle E | \,, \tag{3.4.123}$$

then have a particularly simple time dependence,

$$\begin{aligned}\bigl\langle E,t\bigr| &= \bigl\langle E,t_0\bigr|\,\mathrm{e}^{-\frac{\mathrm{i}}{\hbar}(t-t_0)H} \\ &= \mathrm{e}^{-\frac{\mathrm{i}}{\hbar}(t-t_0)E}\bigl\langle E,t_0\bigr|\,,\end{aligned}\tag{3.4.124}$$

that is: their time dependence is simply an energy-dependent phase factor.

In the present context of a harmonic oscillator, $H=\hbar\omega A^\dagger A$, the energy states are the Fock states, $\bigl\langle E\bigr|=\bigl\langle n\bigr|$ with $E=n\hbar\omega$, and we have

$$\bigl\langle n,t\bigr| = \bigl\langle n,t_0\bigr|\,\mathrm{e}^{-\mathrm{i}\omega(t-t_0)A^\dagger A}\tag{3.4.125}$$

where

$$A^\dagger A = A^\dagger(t)A(t) = A^\dagger(t_0)A(t_0)\tag{3.4.126}$$

does not depend on time. Thus

$$\bigl\langle n,t\bigr| = \mathrm{e}^{-\mathrm{i}n\omega(t-t_0)}\bigl\langle n,t_0\bigr|\,.\tag{3.4.127}$$

We combine it with the equal-time statement

$$\bigl\langle a^*\bigr| = \sum_{n=0}^{\infty}\frac{a^{*n}}{\sqrt{n!}}\bigl\langle n\bigr|\tag{3.4.128}$$

to get

$$\begin{aligned}\bigl\langle a^*,t\bigr| &= \sum_n \frac{a^{*n}}{\sqrt{n!}}\bigl\langle n,t\bigr| \\ &= \sum_n \frac{a^{*n}}{\sqrt{n!}}\,\mathrm{e}^{-\mathrm{i}n\omega(t-t_0)}\bigl\langle n,t_0\bigr| \\ &= \sum_n \frac{\Bigl(a^*\,\mathrm{e}^{-\mathrm{i}\omega(t-t_0)}\Bigr)^n}{\sqrt{n!}}\bigl\langle n,t_0\bigr| \\ &= \bigl\langle a^*\,\mathrm{e}^{-\mathrm{i}\omega(t-t_0)},t_0\bigr|\,,\end{aligned}\tag{3.4.129}$$

thereby reproducing (3.4.118).

In this line of reasoning, we exploit our prior knowledge of the eigenvalues and eigenbras of the Hamilton operator but, in fact, all this information is contained in the time transformation functions $\bigl\langle\dots,t\bigl|\dots,t_0\bigr\rangle$ between some complete set of kets and a complete set of bras. To see the general picture, we consider

$$\bigl\langle\alpha,t\bigl|\beta,t_0\bigr\rangle = \bigl\langle\alpha,t_0\bigr|\,\mathrm{e}^{-\frac{\mathrm{i}}{\hbar}(t-t_0)H}\bigl|\beta,t_0\bigr\rangle\tag{3.4.130}$$

with some general sets of quantum numbers α, β and some time-independent Hamilton operator H. We use the eigenstates of H,

$$H|E\rangle = |E\rangle E, \tag{3.4.131}$$

to write

$$\mathrm{e}^{-\frac{\mathrm{i}}{\hbar}(t-t_0)H} = \sum_E |E,t_0\rangle\, \mathrm{e}^{-\frac{\mathrm{i}}{\hbar}(t-t_0)E} \langle E,t_0| \tag{3.4.132}$$

where $\sum_E$ is a symbolic summation over the eigenvalues of H; it could in fact be an integration. Then

$$\langle \alpha,t|\beta,t_0\rangle = \sum_E \langle \alpha,t_0|E,t_0\rangle\, \mathrm{e}^{-\frac{\mathrm{i}}{\hbar}(t-t_0)E} \langle E,t_0|\beta,t_0\rangle \tag{3.4.133}$$

where we recognize that $\langle \alpha,t_0|E,t_0\rangle = \langle \alpha|E\rangle$ and $\langle E,t_0|\beta,t_0\rangle = \langle E|\beta\rangle$ do not depend on the common time, so that

$$\langle \alpha,t|\beta,t_0\rangle = \sum_E \langle \alpha|E\rangle\, \mathrm{e}^{-\frac{\mathrm{i}}{\hbar}(t-t_0)E} \langle E|\beta\rangle\,. \tag{3.4.134}$$

This tells us that we can extract (i) the eigenvalues E as well as (ii) the probability amplitudes $\langle \alpha|E\rangle$ and $\langle E|\beta\rangle$ by expanding the transformation function $\langle \alpha,t|\beta,t_0\rangle$ into a Fourier sum (or integral) with phase factors of the time difference $t-t_0$. This procedure is quite general.

When applied to

$$\begin{aligned}\langle a^*,t|a',t_0\rangle &= \mathrm{e}^{a^*\,\mathrm{e}^{-\mathrm{i}\omega(t-t_0)}a'}\\ &= \sum_{n=0}^{\infty} \frac{a^{*n}}{\sqrt{n!}}\,\mathrm{e}^{-\mathrm{i}n\omega(t-t_0)}\,\frac{a'^n}{\sqrt{n!}}\end{aligned} \tag{3.4.135}$$

we read off (i) that the eigenvalues of $H = \hbar\omega A^\dagger A$ are $n\hbar\omega$ with $n = 0,1,2,\ldots$, and (ii) that the energy eigenkets $|n\rangle$ and eigenbras $\langle n|$ have equal-time probability amplitudes

$$\langle a^*|n\rangle = \frac{a^{*n}}{\sqrt{n!}}\,, \quad \langle n|a'\rangle = \frac{a'^n}{\sqrt{n!}} \tag{3.4.136}$$

with the eigenbras of $A^\dagger$ and the eigenkets of A, respectively. The symmetric distribution of $1/n!$ is necessary to ensure that these statements are complex conjugates of each other. In short, all the information of (3.4.68) and (3.4.103) is contained in the time transformation function (3.4.115).

Taking yet another step, we note that (3.4.122) and its adjoint statement,

$$\big|\ldots,t\big\rangle = \mathrm{e}^{\frac{\mathrm{i}}{\hbar}(t-t_0)H}\big|\ldots,t_0\big\rangle\,, \tag{3.4.137}$$

imply

$$F(t) = \mathrm{e}^{\frac{\mathrm{i}}{\hbar}(t-t_0)H}F(t_0)\,\mathrm{e}^{-\frac{\mathrm{i}}{\hbar}(t-t_0)H} \tag{3.4.138}$$

for any F that has only a dynamical time dependence but no parametric time dependence, $\partial F/\partial t = 0$. In the present context, the elementary example is $F = A$, for which we know that

$$\begin{aligned} A(t) &= \mathrm{e}^{-\mathrm{i}\omega(t-t_0)}A(t_0) \\ \text{and}\quad A(t) &= \mathrm{e}^{\mathrm{i}\omega(t-t_0)A^\dagger A}A(t_0)\,\mathrm{e}^{-\mathrm{i}\omega(t-t_0)A^\dagger A}\,. \end{aligned} \tag{3.4.139}$$

With $\omega(t-t_0) = \phi$ and all operators at the common time t_0, this says that

$$\mathrm{e}^{-\mathrm{i}\phi A^\dagger A}A\,\mathrm{e}^{\mathrm{i}\phi A^\dagger A} = \mathrm{e}^{\mathrm{i}\phi}A\,. \tag{3.4.140}$$

When presented as

$$A\,\mathrm{e}^{\mathrm{i}\phi A^\dagger A} = \mathrm{e}^{\mathrm{i}\phi(A^\dagger A+1)}A\,, \tag{3.4.141}$$

it illustrates a general relation, namely

$$Af\big(A^\dagger A\big) = f\big(A^\dagger A+1\big)A\,, \tag{3.4.142}$$

which is easily demonstrated by applying the operators to Fock states:

$$\begin{aligned} \big\langle n\big|Af\big(A^\dagger A\big) &= \sqrt{n+1}\big\langle n+1\big|f\big(A^\dagger A\big) \\ &= f(n+1)\sqrt{n+1}\big\langle n+1\big| \\ &= f(n+1)\big\langle n\big|A \\ &= \big\langle n\big|f\big(A^\dagger A+1\big)A\,; \end{aligned} \tag{3.4.143}$$

now multiply by $\big|n\big\rangle$ from the left and sum over n to exploit the completeness relation (3.4.95). The following exercise shows a typical application of (3.4.142).

3-30 Begin with

$$A^\dagger A = f_1(A^\dagger A)\,,$$
$$A^{\dagger 2} A^2 = A^\dagger A(A^\dagger A - 1) = f_2(A^\dagger A)\,,$$

and find a recurrence relation (in m) for

$$A^{\dagger m} A^m = f_m(A^\dagger A) = A^\dagger f_{m-1}(A^\dagger A) A\,.$$

Then conclude that

$$A^{\dagger m} A^m = \frac{(A^\dagger A)!}{(A^\dagger A - m)!}\,.$$

Yet another perspective is to regard $\mathrm{e}^{-\mathrm{i}\phi A^\dagger A} A\, \mathrm{e}^{\mathrm{i}\phi A^\dagger A}$ as defining a ϕ-dependent operator A_ϕ with $A_{\phi=0} = A$. Then differentiate with respect to ϕ,

$$\begin{aligned}
\frac{\partial}{\partial\phi} A_\phi &= \frac{\partial}{\partial\phi}\Bigl(\mathrm{e}^{-\mathrm{i}\phi A^\dagger A} A\, \mathrm{e}^{\mathrm{i}\phi A^\dagger A}\Bigr) \\
&= \mathrm{e}^{-\mathrm{i}\phi A^\dagger A}\, \mathrm{i} \underbrace{[A, A^\dagger A]}_{=A}\, \mathrm{e}^{\mathrm{i}\phi A^\dagger A} \\
&= \mathrm{i}\, \mathrm{e}^{-\mathrm{i}\phi A^\dagger A} A\, \mathrm{e}^{\mathrm{i}\phi A^\dagger A} \\
&= \mathrm{i} A_\phi\,,
\end{aligned} \tag{3.4.144}$$

and solve this differential equation,

$$A_\phi = \mathrm{e}^{\mathrm{i}\phi} A_{\phi=0} = \mathrm{e}^{\mathrm{i}\phi} A\,. \tag{3.4.145}$$

3.5 Two-dimensional harmonic oscillator

3.5.1 *Isotropy*

The Hamilton operator for an isotropic harmonic oscillator in two dimensions is

$$\begin{aligned}
H &= \frac{1}{2M}(P_1^2 + P_2^2) + \frac{1}{2} M\omega^2 (X_1^2 + X_2^2) - \hbar\omega \\
&= \hbar\omega\Bigl({A_1}^\dagger A_1 + {A_2}^\dagger A_2\Bigr)\,,
\end{aligned} \tag{3.5.1}$$

where the isotropy refers to the identical frequencies for both directions. When labeling the eigenkets of H by the oscillator quantum numbers for the two directions, we have

$$\begin{aligned} H\big|n_1,n_2\big\rangle &= \big|n_1,n_2\big\rangle\hbar\omega(n_1+n_2) \\ &= \big|n_1,n_2\big\rangle\hbar\omega N \quad \text{with} \quad N=n_1+n_2 \end{aligned} \tag{3.5.2}$$

so that the energy eigenvalue does not depend on the two quantum numbers individually but only on their sum. As a consequence, there is more than one eigenstate for a given energy. More precisely, the decompositions

$$N = N+0 = (N-1)+1 = (N-2)+2 = \cdots = 0+N \tag{3.5.3}$$

for $N = n_1 + n_2$ show that there are $N+1$ mutually orthogonal states to energy $N\hbar\omega$. Put differently, there is a whole $(N+1)$-dimensional subspace to this energy. Therefore any linear combination of the form

$$\sum_{k=0}^{N}\big|N-k,k\big\rangle\alpha_k$$

with arbitrary complex coefficients α_k is eigenket of H to eigenvalue $N\hbar\omega$.

A systematic *degeneracy* of eigenvalues of this kind is never just an accident, it is always the consequence of a symmetry possessed by the Hamilton operator, sometimes a well hidden symmetry. In the present case, however, it is very clear what that symmmetry is. It is the invariance of H under rotations,

$$\begin{aligned} X_1 \to \overline{X_1} &= X_1\cos\phi + X_2\sin\phi\,, \\ X_2 \to \overline{X_2} &= X_2\cos\phi - X_1\sin\phi\,, \end{aligned} \tag{3.5.4}$$

or, more compactly

$$\begin{pmatrix} X_1 \\ X_2 \end{pmatrix} \to \begin{pmatrix} \overline{X_1} \\ \overline{X_2} \end{pmatrix} = \begin{pmatrix} \cos\phi & \sin\phi \\ -\sin\phi & \cos\phi \end{pmatrix}\begin{pmatrix} X_1 \\ X_2 \end{pmatrix}, \tag{3.5.5}$$

and likewise for the momentum operators

$$\begin{pmatrix} P_1 \\ P_2 \end{pmatrix} \to \begin{pmatrix} \overline{P_1} \\ \overline{P_2} \end{pmatrix} = \begin{pmatrix} \cos\phi & \sin\phi \\ -\sin\phi & \cos\phi \end{pmatrix}\begin{pmatrix} P_1 \\ P_2 \end{pmatrix}. \tag{3.5.6}$$

It is important to note that these transformations are unitary, which we verify by checking that the transformed operators obey the same commutation relations as the original ones. Indeed, the commutators

$$\big[X_j,X_k\big]=0\,,\quad \big[P_j,P_k\big]=0\,,\quad \big[X_j,P_k\big]=\mathrm{i}\hbar\delta_{jk} \tag{3.5.7}$$

for $j,k=1,2$ imply the same relations for the transformed operators,

$$\left[\overline{X_j},\overline{X_k}\right]=0\,,\quad \left[\overline{P_j},\overline{P_k}\right]=0\,,\quad \left[\overline{X_j},\overline{P_k}\right]=\mathrm{i}\hbar\delta_{jk}\,, \tag{3.5.8}$$

as one verifies by inspection. Just one example will suffice as an illustration,

$$\begin{aligned}\left[\overline{X_1},\overline{P_1}\right] &= \left[X_1\cos\phi+X_2\sin\phi, P_1\cos\phi+P_2\sin\phi\right]\\ &= \underbrace{\left[X_1,P_1\right]}_{=\mathrm{i}\hbar}(\cos\phi)^2+\underbrace{\left[X_2,P_2\right]}_{=\mathrm{i}\hbar}(\sin\phi)^2\\ &= \mathrm{i}\hbar\,,\quad \text{indeed}\,.\end{aligned} \tag{3.5.9}$$

And concerning the invariance of H, we note that

$$\begin{aligned}{X_1}^2+{X_2}^2 \to \overline{X_1}^2+\overline{X_2}^2 &= \left(\overline{X_1},\,\overline{X_2}\right)\begin{pmatrix}\overline{X_1}\\ \overline{X_2}\end{pmatrix}\\ &= \left(X_1,\,X_2\right)\underbrace{\begin{pmatrix}\cos\phi & -\sin\phi\\ \sin\phi & \cos\phi\end{pmatrix}\begin{pmatrix}\cos\phi & \sin\phi\\ -\sin\phi & \cos\phi\end{pmatrix}}_{=\begin{pmatrix}1&0\\0&1\end{pmatrix}}\begin{pmatrix}X_1\\ X_2\end{pmatrix}\\ &= \left(X_1,\,X_2\right)\begin{pmatrix}X_1\\ X_2\end{pmatrix}\\ &= {X_1}^2+{X_2}^2\end{aligned} \tag{3.5.10}$$

is invariant, and so are ${P_1}^2+{P_2}^2$ and H itself.

The transformation $X_j,P_k \to \overline{X_j},\overline{P_k}$ does not depend on time as a parameter (but the operators do depend on time, of course) which is to say that we have the same transformation at all times. Therefore, there must be a unitary operator U such that

$$U^\dagger X_j U=\overline{X_j}\,,\quad U^\dagger P_k U=\overline{P_k} \tag{3.5.11}$$

with $\dfrac{\partial}{\partial t}U=0$. We know that H does not change

$$U^\dagger H U = H \tag{3.5.12}$$

so that $HU=UH$ or $[U,H]=0$. This in turn tells us that

$$\frac{\mathrm{d}}{\mathrm{d}t}U=\underbrace{\frac{\partial U}{\partial t}}_{=0}+\underbrace{\frac{1}{\mathrm{i}\hbar}[U,H]}_{=0}=0\,, \tag{3.5.13}$$

and this observation, namely

> If the Hamilton operator is invariant under the unitary transformation effected by U, then U is constant in time.

is an elementary example of a more general statement that is known as Noether's theorem, named after Emmy Noether.

We can think of the transformation $X_j, P_k \to \overline{X_j}, \overline{P_k}$ as coming about in many small steps, small increments of the parameter ϕ, such that finally the total transformation is completed. We write

$$U = \mathrm{e}^{\mathrm{i}\phi G} \tag{3.5.14}$$

for this total transformation, thereby identifying the hermitian generator G of the unitary transformation. It is more systematic to define G by the infinitesimal transformation that increases ϕ by $\delta\phi$. Then

$$\begin{aligned} U &\to (1 + \mathrm{i}\delta\phi G)\,, \\ U^\dagger &\to (1 - \mathrm{i}\delta\phi G) \quad \text{with} \quad G = G^\dagger \end{aligned} \tag{3.5.15}$$

and we have

$$\begin{aligned} \overline{X_j} &= U^\dagger X_j U = (1 - \mathrm{i}\delta\phi G) X_j (1 + \mathrm{i}\delta\phi G) \\ &= X_j - \mathrm{i}\delta\phi [G, X_j] \end{aligned} \tag{3.5.16}$$

and likewise

$$\overline{P_k} = P_k - \mathrm{i}\delta\phi [G, P_k]\,. \tag{3.5.17}$$

Now, for an infinitesimal rotation angle, (3.5.5) and (3.5.6) are

$$\overline{X_1} = X_1 + \delta\phi X_2\,, \qquad \overline{X_2} = X_2 - \delta\phi X_1\,, \tag{3.5.18}$$

and

$$\overline{P_1} = P_1 + \delta\phi P_2\,, \qquad \overline{P_2} = P_2 - \delta\phi P_1\,, \tag{3.5.19}$$

respectively, so that

$$-\mathrm{i}[G, X_1] = X_2\,, \qquad \mathrm{i}[G, X_2] = X_1 \tag{3.5.20}$$

and

$$-\mathrm{i}[G, P_1] = P_2\,, \qquad \mathrm{i}[G, P_2] = P_1\,. \tag{3.5.21}$$

To proceed further we recall the differentiation rules of Exercise 1-15 on page 35,

$$[X, f(X,P)] = \mathrm{i}\hbar \frac{\partial}{\partial P} f(X,P),$$
$$[f(X,P), P] = \mathrm{i}\hbar \frac{\partial}{\partial X} f(X,P), \tag{3.5.22}$$

valid for any operator pair of the position momentum type, that is: for which $[X, P] = \mathrm{i}\hbar$ holds. In the present context we have two such pairs, namely X_1, P_1 and X_2, P_2. Accordingly, the commutator equations (3.5.20) and (3.5.21) can be recast to read

$$-\hbar \frac{\partial}{\partial P_1} G = X_2, \qquad \hbar \frac{\partial}{\partial P_2} G = X_1, \tag{3.5.23}$$

and

$$\hbar \frac{\partial}{\partial X_1} G = P_2, \qquad -\hbar \frac{\partial}{\partial X_1} G = P_2. \tag{3.5.24}$$

The solution is immediate:

$$\hbar G = X_1 P_2 - X_2 P_1, \tag{3.5.25}$$

which is hermitian as it stands since X_1 commutes with P_2 and X_2 with P_1.

We recognize here the third cartesian component (René Descartes) of the angular momentum vector operator $\vec{L} = \vec{R} \times \vec{P}$ in three dimensions

$$\vec{L} = \vec{R} \times \vec{P}: \quad \begin{pmatrix} L_1 \\ L_2 \\ L_3 \end{pmatrix} = \begin{pmatrix} X_1 \\ X_2 \\ X_3 \end{pmatrix} \times \begin{pmatrix} P_1 \\ P_2 \\ P_3 \end{pmatrix} = \begin{pmatrix} X_2P_3 - X_3P_2 \\ X_3P_1 - X_1P_3 \\ X_1P_2 - X_2P_1 \end{pmatrix}, \tag{3.5.26}$$

so $\hbar G = L_3$ and the unitary operator for the *rotation* is

$$U = \mathrm{e}^{\mathrm{i}\phi L_3/\hbar}. \tag{3.5.27}$$

We note the obvious analogy with $\mathrm{e}^{\mathrm{i}xP/\hbar}$, the unitary operator for *translations* that we have met earlier.

3.5.2 *Eigenstates*

We wish to classify the eigenstates of the Hamilton operator in a way that emphasizes the invariance under rotations, that is: we wish to have common eigenkets of H and L_3,

$$H\big|N,m\big\rangle = \big|N,m\big\rangle\hbar\omega N\,,$$
$$L_3\big|N,m\big\rangle = \big|N,m\big\rangle\hbar m\,, \tag{3.5.28}$$

for which purpose we must establish the eigenvalues $\hbar m$ of L_3. We do this by switching to ladder operators, beginning with

$$A_j = \frac{1}{\sqrt{2}}\left(\frac{X_j}{l} + \mathrm{i}\frac{lP_j}{\hbar}\right),$$
$$A_j{}^\dagger = \frac{1}{\sqrt{2}}\left(\frac{X_j}{l} - \mathrm{i}\frac{lP_j}{\hbar}\right), \tag{3.5.29}$$

for which $\big[A_j, A_k{}^\dagger\big] = \delta_{jk}$. Writing

$$X_j = \frac{l}{\sqrt{2}}\big(A_j{}^\dagger + A_j\big)\,, \quad P_j = \frac{\hbar/l}{\sqrt{2}}\big(\mathrm{i}A_j{}^\dagger - \mathrm{i}A_j\big) \tag{3.5.30}$$

we have

$$L_3 = \frac{\hbar}{2}\big(A_1{}^\dagger + A_1\big)\big(\mathrm{i}A_2{}^\dagger - \mathrm{i}A_2\big) - \frac{\hbar}{2}\big(A_2{}^\dagger + A_2\big)\big(\mathrm{i}A_1{}^\dagger - \mathrm{i}A_1\big)$$
$$= \mathrm{i}\hbar\big(A_2{}^\dagger A_1 - A_1{}^\dagger A_2\big)\,, \tag{3.5.31}$$

and we recall that

$$H = \hbar\omega\big(A_1{}^\dagger A_1 + A_2{}^\dagger A_2\big)\,. \tag{3.5.32}$$

It is immediately clear that the Fock states $\big|n_1, n_2\big\rangle$ are not eigenstates of L_3,

$$L_3\big|n_1,n_2\big\rangle = \big|n_1 - 1, n_2 + 1\big\rangle\mathrm{i}\hbar\sqrt{n_1(n_2+1)}$$
$$- \big|n_1 + 1, n_2 - 1\big\rangle\mathrm{i}\hbar\sqrt{(n_1+1)n_2}\,, \tag{3.5.33}$$

where — with the sole exception of $n_1 = n_2 = 0$ — the right-hand side is not a multiple of $\big|n_1, n_2\big\rangle$.

The alternative set of ladder operators defined by

$$A_\pm = \frac{1}{\sqrt{2}}\big(A_1 \mp \mathrm{i}A_2\big)\,, \qquad A_\pm{}^\dagger = \frac{1}{\sqrt{2}}\big(A_1{}^\dagger \pm \mathrm{i}A_2{}^\dagger\big) \tag{3.5.34}$$

are as good as the original ones, because they have analogous commutation relations,

$$\big[A_+, A_+{}^\dagger\big] = 1\,, \qquad \big[A_-, A_-{}^\dagger\big] = 1\,, \qquad \big[A_-, A_+{}^\dagger\big] = 0\,. \tag{3.5.35}$$

Now,

$$A_+{}^\dagger A_+ = \frac{1}{2}\bigl(A_1{}^\dagger A_1 + A_2{}^\dagger A_2 + \mathrm{i}A_2{}^\dagger A_1 - \mathrm{i}A_1{}^\dagger A_2\bigr)\,,$$
$$A_-{}^\dagger A_- = \frac{1}{2}\bigl(A_1{}^\dagger A_1 + A_2{}^\dagger A_2 - \mathrm{i}A_2{}^\dagger A_1 + \mathrm{i}A_1{}^\dagger A_2\bigr) \qquad (3.5.36)$$

are clearly such that

$$A_+{}^\dagger A_+ + A_-{}^\dagger A_- = A_1{}^\dagger A_1 + A_2{}^\dagger A_2$$
$$\text{and} \quad A_+{}^\dagger A_+ - A_-{}^\dagger A_- = \mathrm{i}A_2{}^\dagger A_1 - \mathrm{i}A_1{}^\dagger A_2\,, \qquad (3.5.37)$$

so that

$$H = \hbar\omega\bigl(A_+{}^\dagger A_+ + A_-{}^\dagger A_-\bigr) \qquad (3.5.38)$$

and

$$L_3 = \hbar\bigl(A_+{}^\dagger A_+ - A_-{}^\dagger A_-\bigr)\,. \qquad (3.5.39)$$

Therefore, the common eigenkets of $A_+{}^\dagger A_+$ and $A_-{}^\dagger A_-$,

$$A_+{}^\dagger A_+\bigl|n_+,n_-\bigr\rangle = \bigl|n_+,n_-\bigr\rangle n_+\,,$$
$$A_-{}^\dagger A_-\bigl|n_+,n_-\bigr\rangle = \bigl|n_+,n_-\bigr\rangle n_-\,, \qquad (3.5.40)$$

with $n_\pm = 0,1,2,\ldots$, are also the looked-for joint eigenkets of H and L_3,

$$H\bigl|n_+,n_-\bigr\rangle = \bigl|n_+,n_-\bigr\rangle\hbar\omega(n_+ + n_-)\,,$$
$$L_3\bigl|n_+,n_-\bigr\rangle = \bigl|n_+,n_-\bigr\rangle\hbar(n_+ - n_-)\,. \qquad (3.5.41)$$

Upon identifying

$$\bigl|N,m\bigr\rangle = \bigl|n_+,n_-\bigr\rangle$$
$$\text{with} \quad N = n_+ + n_- \quad \text{and} \quad m = n_+ - n_-\,, \qquad (3.5.42)$$

we have thus found the $\bigl|N,m\bigr\rangle$ kets. For given $N = 0,1,2,\ldots$, the possible values of m are

$$m = N, N-2, \ldots, -N \qquad (3.5.43)$$

a total number of $N+1$ values, as it should be.

This contains an important lesson, perhaps the most significant one of this brief discussion of the isotropic two-dimensional harmonic oscillator. Namely we learn that the eigenvalues of L_3, the third component of the vector operator $\vec{L}$ for orbital angular momentum in (3.5.26), are given by $\hbar m$ with $m = 0, \pm 1, \pm 2, \ldots$, that is

$$\text{the eigenvalues of } L_3 = X_1 P_2 - X_2 P_1 \text{ are } 0, \pm\hbar, \pm 2\hbar, \pm 3\hbar, \ldots . \tag{3.5.44}$$

3-31 Find the response of $A_\pm$ to the rotation $\mathrm{e}^{\mathrm{i}\phi L_3/\hbar}$, that is

$$\mathrm{e}^{-\mathrm{i}\phi L_3/\hbar} A_\pm \, \mathrm{e}^{\mathrm{i}\phi L_3/\hbar} = ?$$

3-32 For $N = 1$ and $N = 2$, express the kets $|n_1, n_2\rangle$ as linear superpositions of the kets $|n_+, n_-\rangle = |N, m\rangle$.

3-33 Express

$$X_\pm = \frac{l}{\sqrt{2}}\left(A_\pm{}^\dagger + A_\pm\right), \qquad P_\pm = \frac{\hbar/l}{\sqrt{2}}\left(\mathrm{i}A_\pm{}^\dagger - \mathrm{i}A_\pm\right)$$

in terms of X_1, X_2, P_1, and P_2. Then use these expressions to find the commutation relations between X_+, X_-, P_+, and P_-. Do you get what you expect?

Chapter 4

Orbital Angular Momentum

4.1 Commutation relations

For the three dimensions of the physical space we have the position vector operator $\vec{R}$ and the momentum vector operator $\vec{P}$ with cartesian components

$$\vec{R} \mathrel{\widehat{=}} \begin{pmatrix} X_1 \\ X_2 \\ X_3 \end{pmatrix}, \qquad \vec{P} \mathrel{\widehat{=}} \begin{pmatrix} P_1 \\ P_2 \\ P_3 \end{pmatrix} \tag{4.1.1}$$

referring to a particular, yet arbitrary, choice of what we call the first, second, and third direction. The fundamental Heisenberg commutation relations

$$\left[X_j, P_k\right] = \mathrm{i}\hbar\delta_{jk} \tag{4.1.2}$$

are compactly summarized by

$$\left[\vec{a} \cdot \vec{R}, \vec{b} \cdot \vec{P}\right] = \mathrm{i}\hbar\, \vec{a} \cdot \vec{b} \tag{4.1.3}$$

where $\vec{a}$ and $\vec{b}$ are *numerical* vectors. We supplement this by the statements that all components of $\vec{R}$ commute with each other,

$$\left[\vec{a} \cdot \vec{R}, \vec{b} \cdot \vec{R}\right] = 0\,, \tag{4.1.4}$$

and the same is true for the components of $\vec{P}$,

$$\left[\vec{a} \cdot \vec{P}, \vec{b} \cdot \vec{P}\right] = 0\,. \tag{4.1.5}$$

The one-dimensional statement

$$[f(X,P),P] = \mathrm{i}\hbar\frac{\partial}{\partial X}f(X,P) \tag{4.1.6}$$

of Exercise 1-15 on page 35 has the three-dimensional generalization

$$\left[f(\vec{R},\vec{P}),\vec{b}\cdot\vec{P}\right] = \mathrm{i}\hbar\,\vec{b}\cdot\frac{\partial}{\partial\vec{R}}f(\vec{R},\vec{P})\,, \tag{4.1.7}$$

where $\frac{\partial}{\partial\vec{R}}$ is the *gradient* differential operator for $\vec{R}$,

$$\frac{\partial}{\partial\vec{R}} \mathrel{\widehat{=}} \begin{pmatrix} \frac{\partial}{\partial X_1} \\ \frac{\partial}{\partial X_2} \\ \frac{\partial}{\partial X_3} \end{pmatrix} \tag{4.1.8}$$

so that $\partial f(\vec{R})/\partial\vec{R}$ is the vector with cartesian components $\partial f/\partial X_1$, $\partial f/\partial X_2$, and $\partial f/\partial X_3$. Likewise we generalize

$$[X,f(X,P)] = \mathrm{i}\hbar\frac{\partial}{\partial P}f(X,P) \tag{4.1.9}$$

to

$$\left[\vec{a}\cdot\vec{R},f(\vec{R},\vec{P})\right] = \mathrm{i}\hbar\,\vec{a}\cdot\frac{\partial}{\partial\vec{P}}f(\vec{R},\vec{P})\,. \tag{4.1.10}$$

The elementary illustration of these differentiation rules is, of course, the Heisenberg commutator itself, inasmuch as

$$\left[\vec{a}\cdot\vec{R},\vec{b}\cdot\vec{P}\right] = \left\{\begin{matrix} \mathrm{i}\hbar\,\vec{b}\cdot\frac{\partial}{\partial\vec{R}}\,\vec{a}\cdot\vec{R} \\ \mathrm{i}\hbar\,\vec{a}\cdot\frac{\partial}{\partial\vec{P}}\,\vec{b}\cdot\vec{P} \end{matrix}\right\} = \mathrm{i}\hbar\,\vec{a}\cdot\vec{b} \tag{4.1.11}$$

exploits the familiar identity

$$\frac{\partial}{\partial\vec{R}}\vec{a}\cdot\vec{R} = \vec{a}\,, \tag{4.1.12}$$

and likewise for the $\vec{P}$ gradient of $\vec{b}\cdot\vec{P}$.

4-1 Consider the hermitian operator $S = \frac{1}{2}(\vec{R}\cdot\vec{P} + \vec{P}\cdot\vec{R})$ and evaluate $\left[\vec{R}, S\right]$ and $\left[S, \vec{P}\right]$. Then find $e^{-i\lambda S}\vec{R}e^{i\lambda S}$ and $e^{-i\lambda S}\vec{P}e^{i\lambda S}$, by first differentiating with respect to the numerical parameter λ.

As noted already at (3.5.26), we further have the *orbital angular momentum* vector operator

$$\vec{L} = \vec{R}\times\vec{P} \tag{4.1.13}$$

with the cartesian components stated there,

$$\begin{aligned} L_1 &= X_2P_3 - X_3P_2\,, \\ L_2 &= X_3P_1 - X_1P_3\,, \\ L_3 &= X_1P_2 - X_2P_1\,. \end{aligned} \tag{4.1.14}$$

We supplement the commutation relations of (3.5.20) and (3.5.21), there stated for $G = L_3/\hbar$,

$$\begin{aligned} [X_1, L_3] &= -i\hbar X_2\,, \qquad & [X_2, L_3] &= i\hbar X_1 \\ [P_1, L_3] &= -i\hbar P_2\,, \qquad & [P_2, L_3] &= i\hbar P_1\,, \end{aligned} \tag{4.1.15}$$

by

$$[X_3, L_3] = 0\,, \qquad [P_3, L_3] = 0 \tag{4.1.16}$$

which say: "rotations around the third axis do not change the third components", and summarize all three compactly in

$$\begin{aligned} \left[\vec{a}\cdot\vec{R}, L_3\right] &= i\hbar(\vec{a}\times\vec{e}_3)\cdot\vec{R}\,, \\ \left[\vec{a}\cdot\vec{P}, L_3\right] &= i\hbar(\vec{a}\times\vec{e}_3)\cdot\vec{P} \end{aligned} \tag{4.1.17}$$

where $\vec{e}_3$ is the unit vector for the third direction, as it appears in the cartesian decomposition

$$\vec{L} = \vec{e}_1L_1 + \vec{e}_2L_2 + \vec{e}_3L_3\,. \tag{4.1.18}$$

With $L_3 = \vec{e}_3\cdot\vec{L}$ on the left, these statements are obviously particular cases of

$$\begin{aligned} \left[\vec{a}\cdot\vec{R}, \vec{b}\cdot\vec{L}\right] &= i\hbar\left(\vec{a}\times\vec{b}\right)\cdot\vec{R}\,, \\ \left[\vec{a}\cdot\vec{P}, \vec{b}\cdot\vec{L}\right] &= i\hbar\left(\vec{a}\times\vec{b}\right)\cdot\vec{P}\,. \end{aligned} \tag{4.1.19}$$

4-2 Alternatively, obtain these commutators by applying (4.1.7) and (4.1.10) to $f(\vec{R},\vec{P}) = \vec{R}\times\vec{P} = \vec{L}$.

Just like $U = \mathrm{e}^{\mathrm{i}\phi L_3/\hbar}$ is the unitary operator for a rotation around the third axis, we have $\mathrm{e}^{\mathrm{i}\phi\vec{e}\cdot\vec{L}/\hbar}$ as the unitary operator for rotations around the axis specified by unit vector $\vec{e}$ and by rotation angle ϕ. For infinitesimal rotations with $\delta\phi\,\vec{e} = \delta\vec{\phi}$, we have

$$\mathrm{e}^{\mathrm{i}\delta\phi\,\vec{e}\cdot\vec{L}/\hbar} = 1 + \frac{\mathrm{i}}{\hbar}\delta\vec{\phi}\cdot\vec{L} \tag{4.1.20}$$

and the effect on the position operator is

$$\begin{aligned}\vec{R} &\to \left(1 - \frac{\mathrm{i}}{\hbar}\delta\vec{\phi}\cdot\vec{L}\right)\vec{R}\left(1 + \frac{\mathrm{i}}{\hbar}\delta\vec{\phi}\cdot\vec{L}\right)\\ &= \vec{R} - \frac{1}{\mathrm{i}\hbar}\left[\vec{R},\delta\vec{\phi}\cdot\vec{L}\right]\\ &= \vec{R} - \delta\vec{\phi}\times\vec{R}\,,\end{aligned} \tag{4.1.21}$$

of which we have seen the $\delta\vec{\phi} \propto \vec{e}_3$ case in (3.5.18) above.

We learn something new by asking the question: What is the difference between performing two consecutive rotations in either order? So, we first rotate by $\delta\vec{\phi}_1$, then by $\delta\vec{\phi}_2$:

$$\begin{aligned}\vec{R} &\to \vec{R} - \delta\vec{\phi}_1\times\vec{R}\\ &\to \vec{R} - \delta\vec{\phi}_2\times\vec{R} - \delta\vec{\phi}_1\times\left(\vec{R} - \delta\vec{\phi}_2\times\vec{R}\right)\\ &= \vec{R} - \left(\delta\vec{\phi}_1 + \delta\vec{\phi}_2\right)\times\vec{R} + \delta\vec{\phi}_1\times\left(\delta\vec{\phi}_2\times\vec{R}\right).\end{aligned} \tag{4.1.22}$$

Or, we first rotate by $\delta\vec{\phi}_2$, then by $\delta\vec{\phi}_1$:

$$\begin{aligned}\vec{R} &\to \vec{R} - \delta\vec{\phi}_2\times\vec{R}\\ &\to \vec{R} - \delta\vec{\phi}_1\times\vec{R} - \delta\vec{\phi}_2\times\left(\vec{R} - \delta\vec{\phi}_1\times\vec{R}\right)\\ &= \vec{R} - \left(\delta\vec{\phi}_1 + \delta\vec{\phi}_2\right)\times\vec{R} + \delta\vec{\phi}_2\times\left(\delta\vec{\phi}_1\times\vec{R}\right).\end{aligned} \tag{4.1.23}$$

The difference

$$\begin{aligned}
&\delta\vec{\phi}_1\times\left(\delta\vec{\phi}_2\times\vec{R}\right)-\delta\vec{\phi}_2\times\left(\delta\vec{\phi}_1\times\vec{R}\right)\\
&\quad=\underbrace{\left[\delta\vec{\phi}_1\times\left(\delta\vec{\phi}_2\times\vec{R}\right)+\delta\vec{\phi}_2\times\left(\vec{R}\times\delta\vec{\phi}_1\right)+\vec{R}\times\left(\delta\vec{\phi}_1\times\delta\vec{\phi}_2\right)\right]}_{=0}\\
&\qquad+\left(\delta\vec{\phi}_1\times\delta\vec{\phi}_2\right)\times\vec{R}\\
&\quad=\left(\delta\vec{\phi}_1\times\delta\vec{\phi}_2\right)\times\vec{R}\\
&\quad=\frac{1}{\mathrm{i}\hbar}\left[\vec{R},\left(\delta\vec{\phi}_1\times\delta\vec{\phi}_2\right)\cdot\vec{L}\right]
\end{aligned}\tag{4.1.24}$$

is itself an infinitesimal rotation. The vanishing of $[\delta\vec{\phi}_1\times\cdots]$ in the second line is an application of Karl Jacobi's identity for vectors. Exercise 3-1 on page 86 in *Basic Matters* deals with the commutator version.

Now, let us look at this same procedure from the point of view of two consecutive unitary transformations. First $\delta\vec{\phi}_1$, then $\delta\vec{\phi}_2$:

$$\begin{aligned}
\vec{R}\to&\left(1-\frac{\mathrm{i}}{\hbar}\delta\vec{\phi}_2\cdot\vec{L}\right)\left(1-\frac{\mathrm{i}}{\hbar}\delta\vec{\phi}_1\cdot\vec{L}\right)\vec{R}\left(1+\frac{\mathrm{i}}{\hbar}\delta\vec{\phi}_1\cdot\vec{L}\right)\left(1+\frac{\mathrm{i}}{\hbar}\delta\vec{\phi}_2\cdot\vec{L}\right)\\
&=\vec{R}-\frac{1}{\mathrm{i}\hbar}\left[\vec{R},\left(\delta\vec{\phi}_1+\delta\vec{\phi}_2\right)\cdot\vec{L}\right]\\
&\quad+\frac{1}{\hbar^2}\left(\delta\vec{\phi}_1\cdot\vec{L}\,\vec{R}\,\delta\vec{\phi}_2\cdot\vec{L}+\delta\vec{\phi}_2\cdot\vec{L}\,\vec{R}\,\delta\vec{\phi}_1\cdot\vec{L}\right)\\
&\quad-\frac{1}{\hbar^2}\left(\vec{R}\,\delta\vec{\phi}_1\cdot\vec{L}\,\delta\vec{\phi}_2\cdot\vec{L}+\delta\vec{\phi}_2\cdot\vec{L}\,\delta\vec{\phi}_1\cdot\vec{L}\,\vec{R}\right);
\end{aligned}\tag{4.1.25}$$

likewise, first $\delta\vec{\phi}_2$, then $\delta\vec{\phi}_1$:

$$\begin{aligned}
\vec{R}\to\vec{R}&-\frac{1}{\mathrm{i}\hbar}\left[\vec{R},\left(\delta\vec{\phi}_1+\delta\vec{\phi}_2\right)\cdot\vec{L}\right]\\
&+\frac{1}{\hbar^2}\left(\delta\vec{\phi}_1\cdot\vec{L}\,\vec{R}\,\delta\vec{\phi}_2\cdot\vec{L}+\delta\vec{\phi}_2\cdot\vec{L}\,\vec{R}\,\delta\vec{\phi}_1\cdot\vec{L}\right)\\
&-\frac{1}{\hbar^2}\left(\vec{R}\,\delta\vec{\phi}_2\cdot\vec{L}\,\delta\vec{\phi}_1\cdot\vec{L}+\delta\vec{\phi}_1\cdot\vec{L}\,\delta\vec{\phi}_2\cdot\vec{L}\,\vec{R}\right).
\end{aligned}\tag{4.1.26}$$

Their difference is

$$\begin{aligned}
&-\frac{1}{\hbar^2}\left[\vec{R},\delta\vec{\phi}_1\cdot\vec{L}\,\delta\vec{\phi}_2\cdot\vec{L}\right]-\frac{1}{\hbar^2}\left[\delta\vec{\phi}_2\cdot\vec{L}\,\delta\vec{\phi}_1\cdot\vec{L},\vec{R}\right]\\
&\qquad=\frac{1}{\mathrm{i}\hbar}\left[\vec{R},\frac{1}{\mathrm{i}\hbar}\left(\delta\vec{\phi}_1\cdot\vec{L}\,\delta\vec{\phi}_2\cdot\vec{L}-\delta\vec{\phi}_2\cdot\vec{L}\,\delta\vec{\phi}_1\cdot\vec{L}\right)\right]\\
&\qquad=\frac{1}{\mathrm{i}\hbar}\left[\vec{R},\frac{1}{\mathrm{i}\hbar}\left[\delta\vec{\phi}_1\cdot\vec{L},\delta\vec{\phi}_2\cdot\vec{L}\right]\right],
\end{aligned}\tag{4.1.27}$$

and the comparison with the difference in (4.1.24) establishes

$$\frac{1}{\mathrm{i}\hbar}\left[\delta\vec{\phi}_1\cdot\vec{L},\delta\vec{\phi}_2\cdot\vec{L}\right]=\left(\delta\vec{\phi}_1\times\delta\vec{\phi}_2\right)\cdot\vec{L}+\boxed{?} \tag{4.1.28}$$

where $\boxed{?}$ commutes with $\vec{R}$, so that it is a function of $\vec{R}$ alone, not containing any component of the momentum operator $\vec{P}$. But, we could have gone through the same chain of arguments by considering successive infinitesimal rotations of $\vec{P}$ rather than $\vec{R}$, and then we would have concluded that $\boxed{?}$ is a function of $\vec{P}$ alone, not containing any components of $\vec{R}$. Therefore, $\boxed{?}$ can at best be a multiple of the identity. It must be proportional to $\delta\vec{\phi}_1$ and proportional to $\delta\vec{\phi}_2$ because the left-hand side has this proportionality. These infinitesimal vectors can only be combined into a number by the scalar product $\delta\vec{\phi}_1\cdot\delta\vec{\phi}_2$, so that

$$\boxed{?}\propto\delta\vec{\phi}_1\cdot\delta\vec{\phi}_2 \tag{4.1.29}$$

follows. Since this is symmetric under the interchange $\delta\vec{\phi}_1\leftrightarrow\delta\vec{\phi}_2$ whereas the left-hand side is antisymmetric, we can match the symmetry only if $\boxed{?}\equiv 0$ altogether. Thus we conclude that

$$\left[\vec{a}\cdot\vec{L},\vec{b}\cdot\vec{L}\right]=\mathrm{i}\hbar\left(\vec{a}\times\vec{b}\right)\cdot\vec{L}\,, \tag{4.1.30}$$

now writing $\vec{a}$ and $\vec{b}$ for $\delta\vec{\phi}_1$ and $\delta\vec{\phi}_2$.

Together with the statements in (4.1.19), we have

$$\left[\vec{a}\cdot\vec{F},\vec{b}\cdot\vec{L}\right]=\mathrm{i}\hbar\left(\vec{a}\times\vec{b}\right)\cdot\vec{F} \tag{4.1.31}$$

where $\vec{F}$ is $\vec{R}$, or $\vec{P}$, or $\vec{L}$, or any linear combination of them, that is: $\vec{F}$ is *any* vector operator,

$$\vec{F}=\alpha\vec{R}+\beta\vec{P}+\gamma\vec{L}\,. \tag{4.1.32}$$

This includes the cases in which α,β,γ are operators themselves, composed of scalar operators such as

$$\vec{R}\cdot\vec{R},\ \vec{R}\cdot\vec{P},\ \vec{P}\cdot\vec{R},\ \vec{L}\cdot\vec{L},\ \ldots$$

because any such dot product of two vectors commutes with $\vec{L}$, as it is invariant under rotations, as you are invited to demonstrate in the following exercise.

4-3 Consider two arbitrary vector operators $\vec{F}_1$ and $\vec{F}_2$. Show that

$$\left[\vec{F}_1 \cdot \vec{F}_2, \vec{L}\right] = 0\,.$$

4.2 Eigenvalues and eigenstates

One such scalar operator is $\vec{L} \cdot \vec{L}$, it commutes with $\vec{L}$,

$$\left[\vec{L} \cdot \vec{L}, \vec{L}\right] = 0\,, \tag{4.2.1}$$

although the components of $\vec{L}$ do not commute among themselves,

$$[L_1, L_2] = \mathrm{i}\hbar L_3\,, \tag{4.2.2}$$

which — we recall — is the essence of the geometrical observation that it matters in which order you perform successive rotations.

As a consequence, there are common eigenstates of $\vec{L}^2 = \vec{L} \cdot \vec{L}$ and either one of its components. As usual we single out the third component and thus look for common eigenstates of $\vec{L}^2$ and L_3,

$$\begin{aligned} \vec{L}^2 \big|l, m\big\rangle &= \big|l, m\big\rangle \hbar^2 l(l+1)\,, \\ L_3 \big|l, m\big\rangle &= \big|l, m\big\rangle \hbar m\,, \end{aligned} \tag{4.2.3}$$

where we already know what is stated in (3.5.44), namely that the possible values for m are $m = 0, \pm 1, \pm 2, \ldots$, all differences $n_+ - n_-$ of (3.5.42). That we write the eigenvalue of $\vec{L}^2$ as $\hbar^2 l(l+1)$ will turn out to be a convenient choice in a short while.

The nonhermitian operators

$$L_\pm = L_1 \pm \mathrm{i} L_2 = {L_\mp}^\dagger \tag{4.2.4}$$

are *ladder operators* for the quantum number m, inasmuch as

$$\begin{aligned} [L_3, L_\pm] &= [L_3, L_1] \pm \mathrm{i}[L_3, L_2] \\ &= \mathrm{i}\hbar L_2 \pm \mathrm{i}(-\mathrm{i}\hbar L_1) \\ &= \pm\hbar(L_1 \pm \mathrm{i} L_2) = \pm\hbar L_\pm \end{aligned} \tag{4.2.5}$$

or

$$L_3 L_\pm = L_\pm (L_3 \pm \hbar)\,. \tag{4.2.6}$$

The ladder operator property is demonstrated by

$$\begin{aligned} L_3 L_\pm \big|l,m\big\rangle &= L_\pm (L_3 \pm \hbar)\big|l,m\big\rangle \\ &= L_\pm \big|l,m\big\rangle \hbar(m\pm 1) \end{aligned} \tag{4.2.7}$$

which implies

$$L_\pm \big|l,m\big\rangle \propto \big|l,m\pm 1\big\rangle \tag{4.2.8}$$

because $L_\pm \big|l,m\big\rangle$ is eigenket of L_3 with eigenvalue $\hbar(m\pm 1)$.

4-4 Why are we sure that, along with the change of m, there is not also a change of l?

As usual, we take the $\big|l,m\big\rangle$ kets as orthonormal,

$$\big\langle l,m\big|l',m'\big\rangle = \begin{Bmatrix} 1 \text{ if } l=l' \text{ and } m=m' \\ 0 \qquad \text{otherwise} \end{Bmatrix} = \delta_{ll'}\delta_{mm'}\,, \tag{4.2.9}$$

and find the proportionality factor in (4.2.8) by evaluating the length of ket $L_\pm\big|l,m\big\rangle$,

$$\big\langle l,m\big|{L_\pm}^\dagger L_\pm\big|l,m\big\rangle = \big\langle l,m\big|L_\mp L_\pm\big|l,m\big\rangle\,. \tag{4.2.10}$$

Now, note that

$$\vec{L}^2 = L_1^2 + L_2^2 + L_3^2 \tag{4.2.11}$$

and

$$\begin{aligned} L_+L_- &= (L_1 + \mathrm{i}L_2)(L_1 - \mathrm{i}L_2) \\ &= L_1^2 + L_2^2 - \mathrm{i}\,[L_1, L_2] \\ &= \vec{L}^2 - L_3^2 + \hbar L_3\,, \\ L_-L_+ &= (L_1 - \mathrm{i}L_2)(L_1 + \mathrm{i}L_2) \\ &= L_1^2 + L_2^2 + \mathrm{i}\,[L_1, L_2] \\ &= \vec{L}^2 - L_3^2 - \hbar L_3\,, \end{aligned} \tag{4.2.12}$$

so that

$$\begin{aligned} \big\langle l,m\big|L_\mp L_\pm\big|l,m\big\rangle &= \big\langle l,m\big|\Big(\vec{L}^2 - L_3^2 \mp \hbar L_3\Big)\big|l,m\big\rangle \\ &= \hbar^2\big(l(l+1) - m^2 \mp m\big) \\ &= \hbar^2(l\mp m)(l\pm m+1)\,. \end{aligned} \tag{4.2.13}$$

This is the squared length of $L_\pm\big|l,m\big\rangle$, and therefore we have

$$L_\pm\big|l,m\big\rangle = \big|l,m\pm 1\big\rangle\hbar\sqrt{(l\mp m)(l\pm m+1)} \tag{4.2.14}$$

with the usual convention that takes the normalization factor positive.

Since

$$\vec{L}^2 - L_3^2 = L_1^2 + L_2^2 \tag{4.2.15}$$

is a positive operator, its eigenvalues cannot be negative, saying that

$$l(l+1) \geq m^2 \tag{4.2.16}$$

for all $\big|l,m\big\rangle$, so the increase of m by one upon application of L_+ must terminate when the largest m value is reached for the given l value,

$$L_+\big|l,m\big\rangle = 0 \quad \text{if } m = \text{maximal value for given } l \text{ value} \tag{4.2.17}$$

so that

$$(l-m)(l+m+1) = 0 \tag{4.2.18}$$

for the largest m value, implying that $m = l$ then. Likewise there must be a minimal m value for which $L_-\big|l,m\big\rangle = 0$ or $(l+m)(l-m+1) = 0$, so that $m = -l$ is that smallest value. In short, we have

$$m = l, l-1, \ldots, -l \tag{4.2.19}$$

for the possible m values to any l. Since we know already that the possible values of m are all integers, $m = 0, \pm 1, \pm 2, \pm 3, \ldots$, it now follows that l can only have the values $l = 0, 1, 2, \ldots$, all nonnegative integers.

In summary, the common eigenkets of $\vec{L}^2$ and L_3 are such that

$$\begin{aligned}\vec{L}^2\big|l,m\big\rangle &= \big|l,m\big\rangle\hbar^2 l(l+1)\,,\\ L_3\big|l,m\big\rangle &= \big|l,m\big\rangle\hbar m\,,\end{aligned} \tag{4.2.20}$$

with $l = 0, 1, 2, \ldots$ and $m = 0, \pm 1, \ldots, \pm l$, so that there are $2l+1$ states for given l. The ladder operators $L_1 \pm \mathrm{i}L_2$ increase or decrease the m value in accordance with

$$\begin{aligned}(L_1 \pm \mathrm{i}L_2)\big|l,m\big\rangle &= \big|l,m\pm 1\big\rangle\hbar\sqrt{(l\mp m)(l\pm m+1)}\,,\\ \big\langle l,m\big|(L_1 \pm \mathrm{i}L_2) &= \hbar\sqrt{(l\pm m)(l\mp m+1)}\big\langle l,m\mp 1\big|\,.\end{aligned} \tag{4.2.21}$$

As a final remark, let us note that we meet a commutation relation of the structure in (4.1.30),

$$\left[\vec{a}\cdot\vec{L},\vec{b}\cdot\vec{L}\right]=\mathrm{i}\hbar\left(\vec{a}\times\vec{b}\right)\cdot\vec{L} \tag{4.2.22}$$

also in (3.5.10) of *Basic Matters* where we have

$$\left[\vec{a}\cdot\vec{\sigma},\vec{b}\cdot\vec{\sigma}\right]=2\mathrm{i}\left(\vec{a}\times\vec{b}\right)\cdot\vec{\sigma}\,. \tag{4.2.23}$$

We multiply by $\left(\frac{1}{2}\hbar\right)^2$,

$$\left[\vec{a}\cdot\frac{\hbar}{2}\vec{\sigma},\vec{b}\cdot\frac{\hbar}{2}\vec{\sigma}\right]=\mathrm{i}\hbar(\vec{a}\times\vec{b})\cdot\frac{\hbar}{2}\vec{\sigma}\,, \tag{4.2.24}$$

and conclude that $\frac{1}{2}\hbar\vec{\sigma}$ has exactly the same commutation relations as the orbital angular momentum $\vec{L}$. But since the eigenvalues of any component of Wolfgang Pauli's vector operator $\vec{\sigma}$ are ±1, we have $\pm\frac{1}{2}\hbar$ for the eigenvalues of $\frac{1}{2}\hbar\sigma_3$, which would mean $m=\pm\frac{1}{2}$. Therefore, $\frac{1}{2}\hbar\vec{\sigma}$ cannot be an orbital angular momentum. Indeed, it is an *intrinsic* angular momentum, called *spin*, for which there is no classical analog. General angular momenta of this kind are discussed in Section 4.1 of *Perturbed Evolution*, but presently we only deal with orbital angular momentum for which l and m are integers, not half-integers.

4-5 Consider an arbitrary vector operator $\vec{F}$ and show that

$$\mathrm{e}^{-\mathrm{i}\phi\vec{e}\cdot\vec{L}/\hbar}\vec{F}\,\mathrm{e}^{\mathrm{i}\phi\vec{e}\cdot\vec{L}/\hbar}=\vec{e}\,\vec{e}\cdot\vec{F}+\vec{e}\times\left(\vec{F}\times\vec{e}\right)\cos\phi-\vec{e}\times\vec{F}\sin\phi$$

where unit vector $\vec{e}$ specifies the axis of rotation, and ϕ is the rotation angle.

4-6 Use $[X_j,P_k]=\mathrm{i}\hbar\,\delta_{jk}$ to verify directly that $[L_1,L_2]=\mathrm{i}\hbar L_3$.

4-7 Show that $\vec{L}\times\vec{L}=\mathrm{i}\hbar\vec{L}$.

4-8 Demonstrate the generalization thereof,

$$\vec{F}\times\vec{L}+\vec{L}\times\vec{F}=2\mathrm{i}\hbar\vec{F}\,,$$

valid for any vector operator $\vec{F}$.

4.3 Differential operators. Polar coordinates

In two dimensions, it is expedient to employ polar coordinates s, ϕ rather than cartesian coordinates x_1, x_2 if the physical system of interest exhibits a rotational symmetry. The two sets of coordinates are related to each other by

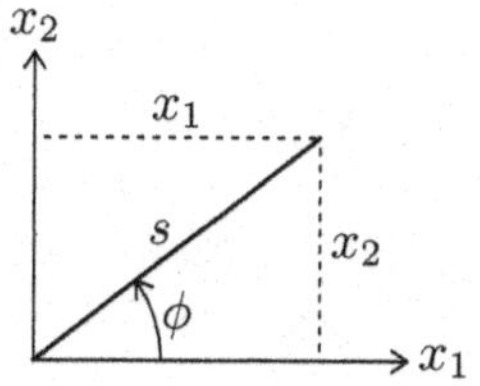

$$\begin{aligned} x_1 &= s\cos\phi\,, \\ x_2 &= s\sin\phi\,, \\ x_1 + \mathrm{i}x_2 &= s\,\mathrm{e}^{\mathrm{i}\phi}\,, \quad s > 0\,, \end{aligned} \tag{4.3.1}$$

and we shall write $|x_1, x_2\rangle = |s, \phi\rangle$ for one and the same eigenket of X_1 and X_2, once labeled by cartesian coordinates, once by polar coordinates, but referring to the same point in the x_1, x_2 plane.

The kinetic energy operator has a corresponding differential operator

$$\langle x_1, x_2|\frac{1}{2M}(P_1^2 + P_2^2) = -\frac{\hbar^2}{2M}\left[\left(\frac{\partial}{\partial x_1}\right)^2 + \left(\frac{\partial}{\partial x_2}\right)^2\right]\langle x_1, x_2|\,, \tag{4.3.2}$$

and so does the angular momentum operator

$$\langle x_1, x_2|L_3 = \frac{\hbar}{\mathrm{i}}\left(x_1\frac{\partial}{\partial x_2} - x_2\frac{\partial}{\partial x_1}\right)\langle x_1, x_2|\,. \tag{4.3.3}$$

We use

$$\begin{aligned} \frac{\partial}{\partial x_1} &= \cos\phi\frac{\partial}{\partial s} - \sin\phi\frac{1}{s}\frac{\partial}{\partial\phi}\,, \\ \frac{\partial}{\partial x_2} &= \sin\phi\frac{\partial}{\partial s} + \cos\phi\frac{1}{s}\frac{\partial}{\partial\phi} \end{aligned} \tag{4.3.4}$$

to express them in polar coordinates, with the outcomes

$$\langle s, \phi|L_3 = \frac{\hbar}{\mathrm{i}}\frac{\partial}{\partial\phi}\langle s, \phi| \tag{4.3.5}$$

and

$$\langle s, \phi|\frac{1}{2M}(P_1^2 + P_2^2) = -\frac{\hbar^2}{2M}\left(\frac{\partial^2}{\partial s^2} + \frac{1}{s}\frac{\partial}{\partial s} + \frac{1}{s^2}\frac{\partial^2}{\partial\phi^2}\right)\langle s, \phi|\,. \tag{4.3.6}$$

The eigenvalue equations (3.5.28) for the common eigenstates of L_3 and the Hamilton operator of the two-dimensional harmonic oscillator,

$$\begin{aligned} L_3\big|N,m\big\rangle &= \big|N,m\big\rangle\hbar m\,, \\ H\big|N,m\big\rangle &= \big|N,m\big\rangle\hbar\omega N\,, \end{aligned} \tag{4.3.7}$$

with the Hamilton operator of (3.5.1),

$$H = \frac{1}{2M}\left(P_1^2+P_2^2\right) + \frac{1}{2}M\omega^2\left(X_1^2+X_2^2\right) - \hbar\omega\,, \tag{4.3.8}$$

are therefore equivalent to the following differential equations for the position wave function $\big\langle x_1,x_2\big|N,m\big\rangle = \big\langle s,\phi\big|N,m\big\rangle$:

$$\begin{aligned} \frac{\hbar}{\mathrm{i}}\frac{\partial}{\partial\phi}\big\langle s,\phi\big|N,m\big\rangle &= \hbar m\big\langle s,\phi\big|N,m\big\rangle\,, \\ \left[-\frac{\hbar^2}{2M}\left(\frac{\partial^2}{\partial s^2}+\frac{1}{s}\frac{\partial}{\partial s}+\frac{1}{s^2}\frac{\partial^2}{\partial\phi^2}\right)+\frac{1}{2}M\omega^2 s^2-\hbar\omega\right]&\big\langle s,\phi\big|N,m\big\rangle \\ &= N\hbar\omega\big\langle s,\phi\big|N,m\big\rangle\,, \end{aligned} \tag{4.3.9}$$

the second of which is the time-independent Schrödinger differential equation. The first equation implies that the ϕ dependence is given by a phase factor $\mathrm{e}^{\mathrm{i}m\phi}$, and so we are invited to write

$$\big\langle s,\phi\big|N,m\big\rangle = \psi_{Nm}(s)\,\mathrm{e}^{\mathrm{i}m\phi}\,. \tag{4.3.10}$$

We insert this into the second equation, note that $\dfrac{\partial}{\partial\phi} \to \mathrm{i}m$ then, and arrive at

$$\left[-\frac{\hbar^2}{2M}\left(\frac{\partial^2}{\partial s^2}+\frac{1}{s}\frac{\partial}{\partial s}\right)+\frac{\hbar^2 m^2}{2Ms^2}+\frac{1}{2}M\omega^2 s^2-(N+1)\hbar\omega\right]\psi_{Nm}(s) = 0\,. \tag{4.3.11}$$

A first simplification results from making use of

$$\frac{\partial^2}{\partial s^2}+\frac{1}{s}\frac{\partial}{\partial s} = \frac{1}{\sqrt{s}}\frac{\partial^2}{\partial s^2}\sqrt{s}+\frac{1}{4s^2}\,, \tag{4.3.12}$$

which gets us to (divide by $\hbar\omega$ as well)

$$\left[-\frac{\hbar}{2M\omega}\frac{\mathrm{d}^2}{\mathrm{d}s^2}+\frac{\hbar}{2M\omega}\frac{m^2-1/4}{s^2}+\frac{1}{2}\frac{M\omega}{\hbar}s^2-(N+1)\right]\sqrt{s}\psi_{Nm}(s) = 0\,. \tag{4.3.13}$$

A further simplification is achieved by switching over to the dimensionless distance variable

$$y = \sqrt{\frac{M\omega}{\hbar}}\,s\,, \qquad \frac{\hbar}{M\omega}\frac{\mathrm{d}^2}{\mathrm{d}s^2} = \frac{\mathrm{d}^2}{\mathrm{d}y^2}\,, \tag{4.3.14}$$

where we recognize once more that $\sqrt{\hbar/(M\omega)}$ is the natural unit of length for a harmonic oscillator. Then (also multiply by -2)

$$\left(\frac{\mathrm{d}^2}{\mathrm{d}y^2} - \frac{m^2-1/4}{y^2} - y^2 + 2N + 2\right)\sqrt{s}\psi_{Nm}(s) = 0\,. \tag{4.3.15}$$

One last step is to realize the possible values for N and m,

$$\begin{aligned} &N = 0,1,2,\ldots \quad \text{and} \quad m = N, N-2, \ldots, -N \\ \text{or} \quad &m = 0,\pm1,\pm2,\ldots \quad \text{and} \quad N = |m|, |m|+2, |m|+4, \ldots\,, \end{aligned} \tag{4.3.16}$$

with the aid of the *radial quantum number* n_r,

$$N = |m| + 2n_r \tag{4.3.17}$$

where now

$$m = 0,\pm1,\pm2,\ldots\,, \qquad n_r = 0,1,2,\ldots \tag{4.3.18}$$

take on their possible values independently of each other. That is, we got rid of the slightly awkward restrictions in (4.3.16) on the values of N or m when the other value is given. With

$$u_{n_r m}(y) = \sqrt{s}\,\psi_{Nm}(s) \tag{4.3.19}$$

we then have

$$\left(\frac{\mathrm{d}^2}{\mathrm{d}y^2} - \frac{m^2-1/4}{y^2} - y^2 + 2|m| + 4n_r + 2\right)u_{n_r m}(y) = 0 \tag{4.3.20}$$

and note in passing that $|m|$, not m is relevant here.

As discussed below, see Section 5.2, the relevant solutions of this differential equation are most easily and compactly expressed in terms of the polynomials that are named after Edmond Laguerre. But right now we are content with having established this differential equation. It will serve an important purpose shortly.

4-9 What can you say about $u_{n_r m}(y)$ for (i) $y \gg 1$ and (ii) $0 < y \ll 1$?

4.4 Differential operators. Spherical coordinates

In three dimensions there is a large choice of particular coordinate systems for special geometries, and one of the most frequently used is the system of spherical coordinates that is particularly well suited for situations with full three-dimensional rotational invariance, that is: full spherical symmetry. This is the case when the Hamilton operator has the form

$$H = \frac{1}{2M}\vec{P}^2 + V(|\vec{R}|) \tag{4.4.1}$$

with a potential energy that *only* depends on the distance $|\vec{R}| = \sqrt{\vec{R}\cdot\vec{R}}$ from the center. Such a Hamilton operator commutes with all components of the orbital angular momentum operator $\vec{L} = \vec{R}\times\vec{P}$,

$$\left[H, \vec{L}\right] = 0\,, \tag{4.4.2}$$

and therefore it is expedient to employ the common eigenstates $|n_r, l, m\rangle$ of $H, \vec{L}^2$, and L_3 as the most natural set of basis states,

$$\begin{aligned} L_3|n_r,l,m\rangle &= |n_r,l,m\rangle\hbar m\,, \\ \vec{L}^2|n_r,l,m\rangle &= |n_r,l,m\rangle\hbar^2 l(l+1)\,, \\ H|n_r,l,m\rangle &= |n_r,l,m\rangle E_{n_r l} \end{aligned} \tag{4.4.3}$$

with the eigenenergies $E_{n_r l}$ depending on the angular momentum quantum number l and the radial quantum number n_r, but not on the quantum number m to L_3.

4-10 Explain why E does not depend on m.

The Schrödinger eigenvalue equation for the wave function $\langle\vec{r}|n_r,l,m\rangle$ follows immediately from

$$\langle\vec{r}|\vec{R} = \vec{r}\langle\vec{r}|\,, \qquad \langle\vec{r}|\vec{P} = \frac{\hbar}{\mathrm{i}}\vec{\nabla}\langle\vec{r}| \tag{4.4.4}$$

and reads $\left(r = |\vec{r}| = \sqrt{\vec{r}\cdot\vec{r}} = \sqrt{x_1^2+x_2^2+x_3^2}\right)$

$$\left[-\frac{\hbar^2}{2M}\vec{\nabla}^2 + V(r)\right]\langle\vec{r}|n_r,l,m\rangle = E_{n_r l}\langle\vec{r}|n_r,l,m\rangle\,. \tag{4.4.5}$$

Since the potential energy part $V(r)$ depends only on the distance r, which is one of the spherical coordinates, we shall use spherical coordinates from here onward. We recall their definition:

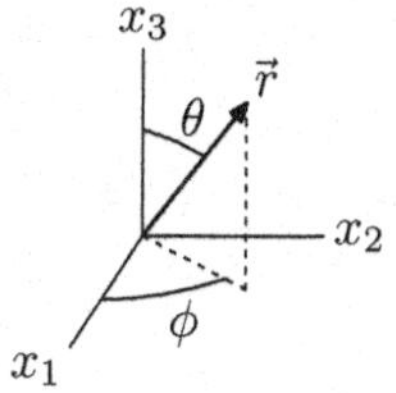

$$
\begin{aligned}
x_1 &= r\sin\theta\,\cos\phi\,,\\
x_2 &= r\sin\theta\,\sin\phi\,,\\
x_3 &= r\cos\theta\,,\quad r>0\,,\ 0\le\theta\le\pi\,,\\
&\qquad\qquad\qquad\qquad 0\le\phi<2\pi\,.
\end{aligned}
\tag{4.4.6}
$$

The local cartesian coordinate system is specified by the unit vector for the r, θ, and ϕ directions

$$
\underbrace{\vec{e}_r \mathrel{\widehat{=}} \begin{pmatrix}\sin\theta\cos\phi\\ \sin\theta\sin\phi\\ \cos\theta\end{pmatrix}}_{\text{pointing “up”}}\,,\quad
\underbrace{\vec{e}_\theta \mathrel{\widehat{=}} \begin{pmatrix}\cos\theta\cos\phi\\ \cos\theta\sin\phi\\ -\sin\theta\end{pmatrix}}_{\text{pointing “south”}}\,,\quad
\underbrace{\vec{e}_\phi \mathrel{\widehat{=}} \begin{pmatrix}-\sin\phi\\ \cos\phi\\ 0\end{pmatrix}}_{\text{pointing “east”}}\,,
\tag{4.4.7}
$$

so that

$$
\begin{aligned}
\vec{r} &= r\vec{e}_r\,,\\
\mathrm{d}\vec{r} &= \vec{e}_r\mathrm{d}r + \vec{e}_\theta\, r\mathrm{d}\theta + \vec{e}_\phi\, r\sin\theta\,\mathrm{d}\phi\,,\\
\text{and}\quad \vec{\nabla} &= \vec{e}_r\frac{\partial}{\partial r} + \vec{e}_\theta\frac{1}{r}\frac{\partial}{\partial\theta} + \vec{e}_\phi\frac{1}{r\sin\theta}\frac{\partial}{\partial\phi}
\end{aligned}
\tag{4.4.8}
$$

are the position vector, its infinitesimal increment, and the gradient differential operator, respectively. We could use them to express $\vec{\nabla}^2$ and $\vec{r}\times\vec{\nabla}$ as differentiations with respect to r, θ, and ϕ.

4-11 Do this to some extent: First get

$$
\vec{r}\times\vec{\nabla} = \vec{e}_\phi\frac{\partial}{\partial\theta} - \vec{e}_\theta\frac{1}{\sin\theta}\frac{\partial}{\partial\phi}
$$

then square it,

$$
\left(\vec{r}\times\vec{\nabla}\right)^2 = \frac{1}{\sin\theta}\frac{\partial}{\partial\theta}\sin\theta\frac{\partial}{\partial\theta} + \frac{1}{(\sin\theta)^2}\frac{\partial^2}{\partial\phi^2}\,.
$$

The result of this exercise would appear in

$$
\begin{aligned}
\langle\vec{r}|\vec{L}^2 = \langle\vec{r}|\left(\vec{R}\times\vec{P}\right)^2 &= \left(\vec{r}\times\frac{\hbar}{\mathrm{i}}\vec{\nabla}\right)^2\langle\vec{r}|\\
&= -\hbar^2\left(\vec{r}\times\vec{\nabla}\right)^2\langle\vec{r}|\,,
\end{aligned}
\tag{4.4.9}
$$

but right now we do not need all the finer details. Rather we note that

$$\vec{L} = \vec{R} \times \vec{P} = -\vec{P} \times \vec{R} \tag{4.4.10}$$

and therefore

$$\vec{L}^2 = -\left(\vec{P} \times \vec{R}\right) \cdot \left(\vec{R} \times \vec{P}\right), \tag{4.4.11}$$

so that

$$\langle\vec{r}|\vec{L}^2 = \hbar^2\left(\vec{\nabla} \times \vec{r}\right) \cdot \left(\vec{r} \times \vec{\nabla}\right)\langle\vec{r}|, \tag{4.4.12}$$

and our attention turns to

$$\begin{aligned}\left(\vec{\nabla} \times \vec{r}\right) \cdot \left(\vec{r} \times \vec{\nabla}\right) &= \vec{\nabla} \cdot \left(\vec{r} \times \left(\vec{r} \times \vec{\nabla}\right)\right) \\ &= \vec{\nabla} \cdot \left(\vec{r}\,\vec{r} \cdot \vec{\nabla} - r^2\vec{\nabla}\right).\end{aligned} \tag{4.4.13}$$

With

$$\vec{\nabla} \cdot \vec{r} = 3 + \vec{r} \cdot \vec{\nabla} \quad \text{and} \quad \vec{\nabla} r^2 = 2\vec{r} + r^2\vec{\nabla} \tag{4.4.14}$$

this differential operator becomes

$$\left(\vec{\nabla} \times \vec{r}\right) \cdot \left(\vec{r} \times \vec{\nabla}\right) = \left(\vec{r} \cdot \vec{\nabla}\right)^2 + \vec{r} \cdot \vec{\nabla} - r^2\vec{\nabla}^2, \tag{4.4.15}$$

and with

$$\vec{r} \cdot \vec{\nabla} = r\frac{\partial}{\partial r} \tag{4.4.16}$$

we arrive at

$$\left(\vec{\nabla} \times \vec{r}\right) \cdot \left(\vec{r} \times \vec{\nabla}\right) = r^2\left(\frac{1}{r}\frac{\partial}{\partial r}r\frac{\partial}{\partial r} + \frac{1}{r}\frac{\partial}{\partial r} - \vec{\nabla}^2\right). \tag{4.4.17}$$

After putting the various pieces together, we have

$$\langle\vec{r}|\vec{L}^2 = \hbar^2 r^2\left(\frac{1}{r}\frac{\partial}{\partial r}r\frac{\partial}{\partial r} + \frac{1}{r}\frac{\partial}{\partial r}\right)\langle\vec{r}| + r^2\langle\vec{r}|\vec{P}^2 \tag{4.4.18}$$

or, upon solving for $\langle\vec{r}|\vec{P}^2$,

$$\langle\vec{r}|\vec{P}^2 = -\hbar^2\left(\frac{1}{r}\frac{\partial}{\partial r}r\frac{\partial}{\partial r} + \frac{1}{r}\frac{\partial}{\partial r}\right)\langle\vec{r}| + \frac{1}{r^2}\langle\vec{r}|\vec{L}^2. \tag{4.4.19}$$

The Schrödinger eigenvalue differential equation (4.4.5) is therefore more explicitly given by

$$-\frac{\hbar^2}{2M}\left(\frac{1}{r}\frac{\partial}{\partial r}r\frac{\partial}{\partial r}+\frac{1}{r}\frac{\partial}{\partial r}\right)\langle\vec{r}|n_r,l,m\rangle+V(r)\langle\vec{r}|n_r,l,m\rangle$$
$$+\frac{1}{2Mr^2}\langle\vec{r}|\underbrace{\vec{L}^2|n_r,l,m\rangle}_{=|n_r,l,m\rangle\hbar^2 l(l+1)}=E_{n_r l}\langle\vec{r}|n_r,l,m\rangle \quad (4.4.20)$$

or

$$\left[-\frac{\hbar^2}{2M}\left(\frac{1}{r}\frac{\partial}{\partial r}r\frac{\partial}{\partial r}+\frac{1}{r}\frac{\partial}{\partial r}\right)+\frac{\hbar^2 l(l+1)}{2Mr^2}+V(r)\right]\langle\vec{r}|n_r,l,m\rangle$$
$$=E_{n_r l}\langle\vec{r}|n_r,l,m\rangle\,. \quad (4.4.21)$$

We know that $L_3 \to \frac{\hbar}{\mathrm{i}}\frac{\partial}{\partial\phi}$, because the ϕ coordinate of spherical coordinates is exactly the same as the ϕ coordinate of polar coordinates, and since L_1 and L_2 are on equal footing with L_3, it follows that they also differentiate an angle dependence, but no radial dependence. As a consequence, the wave function will contain an angular part that is completely determined by the quantum numbers l and m, and a radial part specified by quantum numbers n_r and l,

$$\langle\vec{r}|n_r,l,m\rangle=\psi_{n_r l}(r)Y_{lm}(\theta,\phi)\,, \quad (4.4.22)$$

and the differential equation (4.4.21) is really an equation for $\psi_{n_r l}(r)$ alone.

We also note that

$$\left(\frac{1}{r}\frac{\partial}{\partial r}r\frac{\partial}{\partial r}+\frac{1}{r}\frac{\partial}{\partial r}\right)=\frac{\partial^2}{\partial r^2}+\frac{2}{r}\frac{\partial}{\partial r}=\frac{1}{r}\frac{\partial^2}{\partial r^2}r \quad (4.4.23)$$

and so get to

$$\left[-\frac{\hbar^2}{2M}\frac{\mathrm{d}^2}{\mathrm{d}r^2}+\frac{\hbar^2 l(l+1)}{2Mr^2}+V(r)\right]r\psi_{n_r l}(r)=E_{n_r l}r\psi_{n_r l}(r)\,. \quad (4.4.24)$$

This radial Schrödinger eigenvalue equation looks like a one-dimensional Schrödinger equation with an effective potential energy

$$\frac{\hbar^2 l(l+1)}{2Mr^2}+V(r) \quad (4.4.25)$$

but it must be kept in mind that only $V(r)$ is genuine physical potential energy. The so-called "centrifugal potential" $\frac{\hbar^2 l(l+1)}{2Mr^2}$ is physical kinetic energy. It would be there even in the case of force-free motion.

4-12 Combine the result of Exercise 4-11 on page 121 with another statement in this section and infer that

$$\vec{\nabla}^2 = \frac{1}{r}\frac{\partial^2}{\partial r^2}r + \frac{1}{r^2}\left(\frac{1}{\sin\theta}\frac{\partial}{\partial\theta}\sin\theta\frac{\partial}{\partial\theta} + \frac{1}{(\sin\theta)^2}\frac{\partial^2}{\partial\theta^2}\right)$$

is the spherical-coordinate version of

$$\vec{\nabla}^2 = \frac{\partial^2}{\partial x_1^2} + \frac{\partial^2}{\partial x_2^2} + \frac{\partial^2}{\partial x_3^2},$$

to so-called Laplace differential operator or simply *Laplacian*, named after Marquis de Pierre S. Laplace.

Chapter 5

Hydrogen-like Atoms

5.1 Hamilton operator, Schrödinger equation

For the motion of a single electron — charge $-e$, mass M — in the field of a very heavy nucleus of charge Ze, we have the Hamilton operator

$$H = \frac{1}{2M}\vec{P}^2 - \frac{Ze^2}{|\vec{R}|}, \tag{5.1.1}$$

where we have the attractive Coulomb potential (Charles-Augustin de Coulomb) for the electrostatic interaction between the nuclear charge and the electron charge. For $Z = 1$, this would refer to the hydrogen atom, for $Z = 2$ it is the helium ion He^+, for $Z = 3$, we have the doubly charged Li^{++}, and so forth. The radial Schrödinger eigenvalue equation is then

$$\left[-\frac{\hbar^2}{2M}\frac{\mathrm{d}^2}{\mathrm{d}r^2} + \frac{\hbar^2 l(l+1)}{2Mr^2} - \frac{Ze^2}{r} - E_{n_r l}\right] r\psi_{n_r l}(r) = 0\,. \tag{5.1.2}$$

Just as there is a natural length scale and a natural energy scale for the harmonic oscillator, there is a natural scale for hydrogenic atoms like these, or for atoms in general. It is set by the *Bohr radius*

$$a_0 = \frac{\hbar^2}{Me^2} \quad \text{for the length}\,, \tag{5.1.3}$$

and by

$$\frac{e^2}{a_0} = \frac{Me^4}{\hbar^2} \quad \text{for the energy}\,, \tag{5.1.4}$$

which is twice the so-called *Rydberg constant* Ry, with the names honoring Niels H. D. Bohr and Janne Ryberg, respectively. Their numerical values are

$$a_0 = 0.5292\,\text{Å}\,, \qquad \frac{e^2}{a_0} = 2\,\text{Ry} = 27.2\,\text{eV}\,, \tag{5.1.5}$$

which remind us of the fact that atoms are very tiny objects.

So we write

$$r = a_0 y\,, \qquad E_{n_r l} = \frac{e^2}{a_0}\mathcal{E}_{n_r l}\,, \tag{5.1.6}$$

where y and $\mathcal{E}_{n_r l}$ are dimensionless, and get

$$\left[\frac{\mathrm{d}^2}{\mathrm{d}y^2} - \frac{l(l+1)}{y^2} + \frac{2Z}{y} + 2\mathcal{E}_{n_r l}\right] u_{n_r l}(y) = 0 \tag{5.1.7}$$

with $u_{n_r l}(y) = r\psi_{n_r l}(r)$.

This is, in fact, not an entirely new equation for us, as will become obvious after the change of variables from y to x in accordance with

$$2\lambda y = x^2\,, \qquad \lambda \mathrm{d}y = x\mathrm{d}x\,, \qquad \frac{\mathrm{d}}{\mathrm{d}y} = \frac{\lambda}{x}\frac{\mathrm{d}}{\mathrm{d}x} \tag{5.1.8}$$

where λ is a parameter to be fixed later. Now

$$\begin{aligned} \left(\frac{\mathrm{d}}{\mathrm{d}y}\right)^2 &= \left(\frac{\lambda}{x}\frac{\mathrm{d}}{\mathrm{d}x}\right)^2 = \lambda^2 \frac{1}{x}\frac{\mathrm{d}}{\mathrm{d}x}\frac{1}{x}\frac{\mathrm{d}}{\mathrm{d}x} = \frac{\lambda^2}{x^2}\left[\left(\frac{\mathrm{d}}{\mathrm{d}x}\right)^2 - \frac{1}{x}\frac{\mathrm{d}}{\mathrm{d}x}\right] \\ &= \frac{\lambda^2}{x^2}\sqrt{x}\left(\frac{\mathrm{d}^2}{\mathrm{d}x^2} - \frac{3}{4x^2}\right)\frac{1}{\sqrt{x}}\,, \end{aligned} \tag{5.1.9}$$

and the differential equation for $u_{n_r l}(y)$ reads

$$\frac{\lambda^2}{\sqrt{x^3}}\left[\frac{\mathrm{d}^2}{\mathrm{d}x^2} - \frac{3}{4x^2} - \frac{l(l+1)}{y^2}\frac{x^2}{\lambda^2} + \frac{2Z}{y}\frac{x^2}{\lambda^2} + 2\mathcal{E}\frac{x^2}{\lambda^2}\right]\frac{u(y)}{\sqrt{x}} = 0\,, \tag{5.1.10}$$

where

$$\frac{l(l+1)}{y^2}\frac{x^2}{\lambda^2} = \frac{4l(l+1)}{x^2} = \frac{(2l+1)^2-1}{x^2} \quad \text{and} \quad \frac{2Z}{y}\frac{x^2}{\lambda^2} = \frac{4Z}{\lambda}\,, \tag{5.1.11}$$

so that

$$\left[\frac{\mathrm{d}^2}{\mathrm{d}x^2} - \frac{(2l+1)^2 - \frac{1}{4}}{x^2} + \frac{4Z}{\lambda} + \frac{2\mathcal{E}}{\lambda^2}x^2\right]\frac{u(y)}{\sqrt{x}} = 0\,. \tag{5.1.12}$$

Compare this with (4.3.20), the radial Schrödinger equation for the two-dimensional harmonic oscillator,

$$\left[\frac{\mathrm{d}^2}{\mathrm{d}y^2} - \frac{m^2 - 1/4}{y^2} - y^2 + 2|m| + 4n_r + 2\right]u(y) = 0\,, \tag{5.1.13}$$

and recognize that the two equations are actually the same provided that we use this "dictionary" for the translation:

two-dim. oscillator	three-dim. Coulomb
$m^2 - 1/4$	$(2l+1)^2 - 1/4$
-1	$2\mathcal{E}/\lambda^2$
$2\|m\| + 4n_r + 2$	$4Z/\lambda$

This tells us that the Coulombic eigenvalues are

$$\mathcal{E} = -\frac{1}{2}\lambda^2 \quad \text{with} \quad \lambda = \frac{2Z}{|m| + 2n_r + 1} \quad \text{where } |m| = 2l+1 \tag{5.1.14}$$

that is

$$\mathcal{E}_{n_r l} = -\frac{1}{2}\left(\frac{2Z}{2l+1+2n_r+1}\right)^2 = -\frac{Z^2/2}{(n_r+l+1)^2} \tag{5.1.15}$$

or

$$E_{n_r l} = -\frac{Z^2}{2n^2}\frac{e^2}{a_0} \tag{5.1.16}$$

with the so-called *principal quantum number*

$$n = n_r + l + 1\,. \tag{5.1.17}$$

Inasmuch as $n_r = 0, 1, 2, \ldots$ and $l = 0, 1, 2, \ldots$ independently, we have

$$n = 1, 2, 3, \ldots \tag{5.1.18}$$

and there are a total number of

$$\sum_{l=0}^{n-1}\sum_{m=-l}^{l} 1 = \sum_{l=0}^{n-1}(2l+1) = n^2 \tag{5.1.19}$$

(orbital) states for given n. In summary, the eigenstates of hydrogen-like atoms are such that

$$\begin{aligned} L_3\big|n_r,l,m\big\rangle &= \big|n_r,l,m\big\rangle\hbar m\,, \\ \vec{L}^2\big|n_r,l,m\big\rangle &= \big|n_r,l,m\big\rangle\hbar^2 l(l+1)\,, \\ H\big|n_r,l,m\big\rangle &= \big|n_r,l,m\big\rangle\left(-\frac{Z^2}{2n^2}\frac{e^2}{a_0}\right) \end{aligned} \tag{5.1.20}$$

with $n = n_r + l + 1$. Since the energy eigenvalues are negative, these are bound states in the Coulomb potential. The scattering states, for which H has a positive eigenvalue, cannot be found by such an analogy with the two-dimensional oscillator. They need more mathematical machinery than we have at our disposal.

Rotational symmetry alone accounts for a multiplicity of $2l + 1$ states for each n_r, l pair because the energy cannot depend on the L_3 value, that is on quantum number $m = l, l-1, \ldots, -l$ which takes on $2l + 1$ different values. In the Coulomb potential, there is, however, an additional degeneracy because the eigenenergies do not depend on n_r and l individually, but only on their sum $n_r + l = n - 1$. This indicates that there is an additional dynamical symmetry that is not purely geometrical. We infer that there must therefore also exist another conserved quantity, the generator of the transformation under which H in invariant. Indeed, there is such a conserved quantity, it is the vector

$$\begin{aligned} \vec{A} &= \frac{\vec{R}}{|\vec{R}|} - \frac{1}{2MZe^2}\left(\vec{P}\times\vec{L} - \vec{L}\times\vec{P}\right) \\ &= \frac{\vec{R}}{|\vec{R}|} - \frac{a_0/\hbar^2}{2Z}\left(\vec{P}\times\vec{L} - \vec{L}\times\vec{P}\right). \end{aligned} \tag{5.1.21}$$

One name for $\vec{A}$ is *axis vector*, which refers to its very simple geometrical meaning in classical physics, where the orbits are the familiar ellipses that Johannes Kepler discovered:

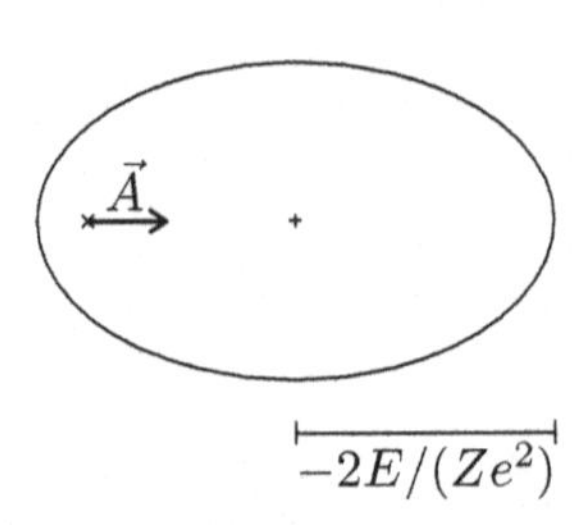

Vector $\vec{A}$ points from the center of the force, located at a focus of the ellipse, to the center of the ellipse, and the length A of $\vec{A}$ is the eccentricity of the ellipse. For total energy E, the major semiaxis of the ellipse is given by $-2E/(Ze^2)$, and the distance from focus to center is A times the major semiaxis.

The axis vector is conserved for the Coulomb potential because the orbits are *closed*, which is an exceptional situation. In general, the orbits in a central force potential $V(r)$ are not closed in classical mechanics, and then we do not get an additional systematic degeneracy of the eigenvalues of the corresponding quantum system.

5.2 Wave functions

5.2.1 *Two-dimensional harmonic oscillator*

For the two-dimensional harmonic oscillator, we know the wave functions for states with definite energy if they are labeled by the quantum numbers n_1 and n_2 that refer to the cartesian coordinates x_1, x_2. We summarize this knowledge in

$$\begin{aligned}\langle x_1, x_2|a_1, a_2\rangle &= \langle x_1|a_1\rangle\langle x_2|a_2\rangle \\ &= \left(\pi^{-1/4}l^{-1/2}\right)^2 \mathrm{e}^{-\frac{x_1^2+x_2^2}{2l^2}}\,\mathrm{e}^{\sqrt{2}\frac{x_1a_1+x_2a_2}{l}}\,\mathrm{e}^{-\frac{1}{2}(a_1^2+a_2^2)} \\ &= \sum_{n_1=0}^{\infty}\sum_{n_2=0}^{\infty}\langle x_1, x_2|n_1, n_2\rangle\langle n_1, n_2|a_1, a_2\rangle \\ &= \sum_{n_1=0}^{\infty}\sum_{n_2=0}^{\infty}\langle x_1, x_2|n_1, n_2\rangle\frac{a_1^{n_1}}{\sqrt{n_1!}}\frac{a_2^{n_2}}{\sqrt{n_2!}}\,, \end{aligned} \tag{5.2.1}$$

which uses the position wave function for coherent states twice, and the expansion in powers of a_1 and a_2 can be performed with the aid of the generating functions for the Hermite polynomials, doubling the effort of Exercise 3-24 on page 92.

If we wish, however, to find the wave functions to the eigenkets $|N, m\rangle = |n_+, n_-\rangle$, the common eigenkets of H and L_3 in (3.5.41), then we should use the coherent states for $A_\pm = \frac{1}{\sqrt{2}}(A_1 \mp \mathrm{i}A_2)$ instead, and we expect that they fit much better to polar coordinates than to cartesian coordinates. In fact, we know already that $\langle s, \phi|N, m\rangle\,\mathrm{e}^{-\mathrm{i}m\phi}$ is a function of s alone, where we continue to identify

$$\begin{aligned}\big\langle s,\phi\big| &= \big\langle x_1,x_2\big| \quad \text{for} \quad x_1+\mathrm{i}x_2 = s\,\mathrm{e}^{\mathrm{i}\phi} \quad \text{with} \quad s>0\,,\\ \big|N,m\big\rangle &= \big|n_+,n_-\big\rangle \quad \text{for} \quad N=n_++n_- \quad \text{and} \quad m=n_+-n_-\,. \end{aligned} \tag{5.2.2}$$

We saw in Section 4.3 that it is convenient to introduce the radial quantum number n_r in accordance with

$$N = |m| + 2n_r \tag{5.2.3}$$

or

$$\begin{aligned} n_r &= \frac{1}{2}(N-|m|) = \frac{1}{2}(n_++n_--|n_+-n_-|)\\ &= \mathrm{Min}\{n_+,n_-\} \equiv n_<\,, \end{aligned} \tag{5.2.4}$$

that is: n_r is the smaller one of n_+ and n_-. The larger one of them is

$$\begin{aligned} n_> \equiv \mathrm{Max}\{n_+,n_-\} &= \frac{1}{2}(n_++n_-+|n_+-n_-|)\\ &= \frac{1}{2}(N+|m|) = n_r+|m| = n_<+|m|\,. \end{aligned} \tag{5.2.5}$$

Equivalent ways of labeling the common eigenkets of H and L_3 are thus

$$\big|N,m\big\rangle = \big|n_r,m\big\rangle = \big|n_+,n_-\big\rangle\,, \tag{5.2.6}$$

of which the middle version is perhaps the most natural one.

In view of (3.5.34),

$$A_\pm = \frac{1}{\sqrt{2}}(A_1 \mp \mathrm{i}A_2)\,, \tag{5.2.7}$$

we have a corresponding choice for the labeling of the coherent states,

$$\big|a_1,a_2\big\rangle = \big|a_+,a_-\big\rangle \quad \text{with} \quad a_\pm = \frac{1}{\sqrt{2}}(a_1\mp \mathrm{i}a_2)\,. \tag{5.2.8}$$

Then

$$\frac{1}{2}\left(a_1^2+a_2^2\right) = \frac{1}{\sqrt{2}}(a_1-\mathrm{i}a_2)\frac{1}{\sqrt{2}}(a_1+\mathrm{i}a_2) = a_+a_- \tag{5.2.9}$$

and

$$\begin{aligned}\sqrt{2}(x_1a_1+x_2a_2) &= x_1(a_++a_-)+x_2(\mathrm{i}a_+-\mathrm{i}a_-)\\ &= (x_1+\mathrm{i}x_2)a_+ + (x_1-\mathrm{i}x_2)a_-\\ &= s\,\mathrm{e}^{\mathrm{i}\phi}a_+ + s\,\mathrm{e}^{-\mathrm{i}\phi}a_-\,, \end{aligned} \tag{5.2.10}$$

and with $x_1^2 + x_2^2 = s^2$ we have therefore

$$\begin{aligned}\langle s, \phi | a_+, a_- \rangle &= \frac{1}{\sqrt{\pi} l} \mathrm{e}^{-\frac{1}{2}(s/l)^2 + (s/l)(\mathrm{e}^{\mathrm{i}\phi} a_+ + \mathrm{e}^{-\mathrm{i}\phi} a_-) - a_+ a_-} \\ &= \sum_{n_+, n_- = 0}^{\infty} \langle s, \phi | n_+, n_- \rangle \frac{a_+^{n_+} a_-^{n_-}}{\sqrt{n_+! n_-!}} . \end{aligned} \tag{5.2.11}$$

We are thus asked to expand the exponential functions in powers of a_+ and a_-, which we do with the aid of this generating function

$$\mathrm{e}^{-xy + \alpha x + \beta y} = \sum_{j=0}^{\infty} \sum_{k=0}^{\infty} \frac{x^j y^k}{j!} (-1)^k \alpha^{j-k} \mathrm{L}_k^{(j-k)}(\alpha\beta) \tag{5.2.12}$$

for the *Laguerre polynomials* that can be defined by their Rodrigues formula (Benjamin O. Rodrigues),

$$\mathrm{L}_n^{(\alpha)}(x) = \frac{1}{n!} x^{-\alpha} \mathrm{e}^{x} \left(\frac{\mathrm{d}}{\mathrm{d}x} \right)^n x^{n+\alpha} \mathrm{e}^{-x} , \tag{5.2.13}$$

for example. One calls $\mathrm{L}_n^{(\alpha)}$ the Laguerre polynomial of degree n and index α. The index can be any complex number, but usually one prefers it to be real and larger than -1.

In the generating function (5.2.12), the index takes on all integer values, positive and negative, but in fact we can always have it nonnegative when picking out particular powers of x and y, because the left-hand side is invariant under the joint interchange of $x \leftrightarrow y$ and $\alpha \leftrightarrow \beta$, so that

$$\mathrm{e}^{-xy + \alpha x + \beta y} = \sum_{j=0}^{\infty} \sum_{k=0}^{\infty} \frac{y^j x^k}{j!} (-1)^k \beta^{j-k} \mathrm{L}_k^{(j-k)}(\alpha\beta) \tag{5.2.14}$$

is an equivalent way of expanding in powers of x and y.

In our application, we thus have a choice between

$$\begin{aligned} \alpha = \beta = \frac{s}{l} \quad \text{and} \quad &\text{(i)} \quad x = \mathrm{e}^{\mathrm{i}\phi} a_+ , \quad && y = \mathrm{e}^{-\mathrm{i}\phi} a_- \\ & \qquad j = n_+ , && k = n_- ; \\ \text{or} \quad &\text{(ii)} \quad x = \mathrm{e}^{-\mathrm{i}\phi} a_- , && y = \mathrm{e}^{\mathrm{i}\phi} a_+ \\ & \qquad j = n_- , && k = n_+ ; \end{aligned} \tag{5.2.15}$$

both to be used in (5.2.12), the first version of the generating function.

Accordingly,

$$
\begin{aligned}
\langle s,\phi|a_+,a_-\rangle\sqrt{\pi}\,l\,\mathrm{e}^{\frac{1}{2}(s/l)^2} &= \mathrm{e}^{-a_+a_- + (s/l)(a_+\,\mathrm{e}^{\mathrm{i}\phi} + a_-\,\mathrm{e}^{-\mathrm{i}\phi})} \\
&= \sum_{n_+,n_-=0}^{\infty} \frac{\left(a_+\,\mathrm{e}^{\mathrm{i}\phi}\right)^{n_+}\left(a_-\,\mathrm{e}^{-\mathrm{i}\phi}\right)^{n_-}}{n_+!}(-1)^{n_-}\left(\frac{s}{l}\right)^{n_+-n_-}\mathrm{L}_{n_-}^{(n_+-n_-)}\left(\left(\frac{s}{l}\right)^2\right) \\
&= \sum_{n_+,n_-=0}^{\infty} \frac{\left(a_-\,\mathrm{e}^{-\mathrm{i}\phi}\right)^{n_-}\left(a_+\,\mathrm{e}^{\mathrm{i}\phi}\right)^{n_+}}{n_-!}(-1)^{n_+}\left(\frac{s}{l}\right)^{n_--n_+}\mathrm{L}_{n_+}^{(n_--n_+)}\left(\left(\frac{s}{l}\right)^2\right),
\end{aligned}
\tag{5.2.16}
$$

where the ϕ dependence is given by

$$
\left(\mathrm{e}^{\mathrm{i}\phi}\right)^{n_+-n_-} = \mathrm{e}^{\mathrm{i}m\phi} \tag{5.2.17}
$$

in both versions, as it should be. When picking out the term proportional to $a_+^{n_+}a_-^{n_-}$ as required by

$$
\langle s,\phi|a_+,a_-\rangle = \sum_{n_+,n_-}\langle s,\phi|n_+,n_-\rangle\frac{a_+^{n_+}a_-^{n_-}}{\sqrt{n_+!\,n_-!}}, \tag{5.2.18}
$$

we exploit the choice between the two versions in (5.2.16) such that the index of the Laguerre polynomial is nonnegative. That is to say, we use the upper version for $n_+ > n_-$, and the lower version for $n_+ < n_-$, and it does not matter which one if $n_+ = n_-$. With this convention, we have

$$
\sum_{n_+,n_-=0}^{\infty}\frac{a_+^{n_+}a_-^{n_-}}{n_>!}\,\mathrm{e}^{\mathrm{i}(n_+\,-\,n_-)\phi}(-1)^{n_<}\left(\frac{s}{l}\right)^{n_>-n_<}\mathrm{L}_{n_<}^{(n_>-n_<)}\left(\left(\frac{s}{l}\right)^2\right) \tag{5.2.19}
$$

as a common way of writing the two series in (5.2.16). As a consequence,

$$
\begin{aligned}
\langle s,\phi|n_+,n_-\rangle = {}& \frac{1}{\sqrt{\pi}l}\,\mathrm{e}^{-\frac{1}{2}(s/l)^2}\sqrt{\frac{n_<!}{n_>!}}\,\mathrm{e}^{\mathrm{i}(n_+\,-\,n_-)\phi} \\
&\times(-1)^{n_<}\left(\frac{s}{l}\right)^{n_>-n_<}\mathrm{L}_{n_<}^{(n_>-n_<)}\left(\left(\frac{s}{l}\right)^2\right)
\end{aligned}
\tag{5.2.20}
$$

or, with $n_+ - n_- = m$, $n_> - n_< = |m|$, $n_< = n_r$, and $n_> = n_r + |m|$,

$$\langle s, \phi | n_r, m \rangle = \frac{\mathrm{e}^{\mathrm{i}m\phi}}{\sqrt{2\pi}} \frac{(-1)^{n_r}}{l} \sqrt{2 \frac{n_r!}{(n_r + |m|)!}} \times \left(\frac{s}{l}\right)^{|m|} \mathrm{L}_{n_r}^{(|m|)}\left(\left(\frac{s}{l}\right)^2\right) \mathrm{e}^{-\frac{1}{2}\left(\frac{s}{l}\right)^2} . \tag{5.2.21}$$

The convention of having a factor $1/\sqrt{2\pi}$ with $\mathrm{e}^{\mathrm{i}m\phi}$ originates in the observation that then the ϕ parts of the wave functions, the azimuthal wave functions, are orthonormal by themselves,

$$\int_0^{2\pi} \mathrm{d}\phi \, \frac{\mathrm{e}^{-\mathrm{i}m'\phi}}{\sqrt{2\pi}} \frac{\mathrm{e}^{\mathrm{i}m\phi}}{\sqrt{2\pi}} = \delta_{mm'} . \tag{5.2.22}$$

And the radial wave functions $S_{n_r,|m|}(s)$,

$$\langle s, \phi | n_r, m \rangle = \frac{\mathrm{e}^{\mathrm{i}m\phi}}{\sqrt{2\pi}} S_{n_r,|m|}(s) , \tag{5.2.23}$$

are then such that

$$\int_0^{\infty} \mathrm{d}s \, s \, S_{n_r,|m|}(s) S_{n'_r,|m|}(s) = \delta_{n_r n'_r} \tag{5.2.24}$$

states their orthonormality.

5-1 Get this from $\langle n_r, m | n'_r, m' \rangle = \delta_{n_r n'_r} \delta_{m,m'}$.

5.2.2 *Hydrogenic atoms*

What is called $S_{n_r,|m|}(s)$ here differs from $\psi_{Nm}(s)$ in (4.3.10) by just the $1/\sqrt{2\pi}$ factor that we choose to associate with the ϕ dependence. Therefore, $\sqrt{s} S_{n_r,|m|}(s)$ obeys the differential equation (4.3.15) for $\sqrt{s}\psi_{Nm}(s)$, where

$$y = \sqrt{\frac{M\omega}{\hbar}} s = \frac{s}{l} . \tag{5.2.25}$$

The substitution of (5.1.8), namely

$$\left(\frac{s}{l}\right)^2 \to 2\lambda y \quad \text{with} \quad y = \frac{r}{a_0} \quad \text{and} \quad \lambda = \frac{Z}{n} \tag{5.2.26}$$

turns it into the radial part of the corresponding wave function for the Coulomb potential. With proper normalization, this eventually gives

$$\langle \vec{r}|n,l,m\rangle = R_{nl}(r)Y_{lm}(\theta,\phi) \tag{5.2.27}$$

with the radial wave functions

$$\begin{aligned} R_{nl}(r) &= (-1)^{n-l-1}\frac{2}{n^2}\left(\frac{Z}{a_0}\right)^{3/2}\sqrt{\frac{(n-l-1)!}{(n+l)!}}\left(\frac{2Zr}{na_0}\right)^l \\ &\quad\times \mathrm{L}_{n-l-1}^{(2l+1)}\left(\frac{2Zr}{na_0}\right)\mathrm{e}^{-\frac{Zr}{na_0}}\,, \end{aligned} \tag{5.2.28}$$

where we have chosen the usual phase conventions.

The orthonormality relation for the radial wave functions is here

$$\int_0^\infty \mathrm{d}r\, r^2\, R_{nl}(r)R_{n'l}(r) = \delta_{nn'} \tag{5.2.29}$$

and one refers to

$$D_{nl}(r) \equiv r^2|R_{nl}(r)|^2 \tag{5.2.30}$$

as the *radial density*, a term suggested by the fact that

$$\int_0^\infty \mathrm{d}r\, D_{nl}(r) = 1\,. \tag{5.2.31}$$

5-2 Use the definition of $\mathrm{L}_n^{(\alpha)}(x)$ in (5.2.13) to find the Laguerre polynomials that you need for $D_{10}(r)$, $D_{20}(r)$, and $D_{21}(r)$. Then employ your favorite software to produce plots of

$$\frac{a_0}{Z}D_{nl}\left(\frac{a_0}{Z}x\right)$$

as functions of x, for $n=1,\ l=0$ and $n=2,\ l=0,1$.

The angular part of the hydrogenic wave functions is given by the factor $Y_{lm}(\theta,\phi)$, which are the familiar *spherical harmonics*, a term coined by William Thomson (Lord Kelvin). They are defined by

$$Y_{lm}(\theta,\phi) = \frac{\mathrm{e}^{im\phi}}{\sqrt{2\pi}}\Theta_{lm}(\theta) \tag{5.2.32}$$

where

$$\Theta_{lm}(\theta) = \sqrt{\frac{2l+1}{2}\frac{(l+m)!}{(l-m)!}}(\sin\theta)^{-m}\left(\frac{\mathrm{d}}{\mathrm{d}\cos\theta}\right)^{l-m}\frac{\left(\cos^2\theta - 1\right)^l}{2^l l!} \tag{5.2.33}$$

can be expressed as an *associated Legendre function*, named after Adrien M. Legendre. The Legendre functions are closely related to the *Legendre polynomials* $\mathrm{P}_l(x)$ that can be defined by their generating function,

$$\frac{1}{\sqrt{1-2tx+t^2}} = \sum_{l=0}^{\infty} t^l \mathrm{P}_l(x)\,, \tag{5.2.34}$$

for example. For further details, please consult any text on mathematical methods for physicists, where special functions like these are treated. We shall be content with noting the orthonormality relation

$$\int_0^{2\pi} \mathrm{d}\phi \int_0^{\pi} \mathrm{d}\theta\, \sin\theta\, Y_{lm}(\theta,\phi)^* Y_{l'm'}(\theta,\phi) = \delta_{ll'}\delta_{mm'}\,. \tag{5.2.35}$$

Harkening back to Exercise 4-11 on page 121, we observe that Y_{lm} obeys the differential equation

$$-\left[\frac{1}{\sin\theta}\frac{\partial}{\partial\theta}\sin\theta\frac{\partial}{\partial\theta} + \left(\frac{1}{\sin\theta}\right)^2\frac{\partial^2}{\partial\phi^2}\right] Y_{lm}(\theta,\phi) = l(l+1)Y_{lm}(\theta,\phi)\,, \tag{5.2.36}$$

which is the eigenvalue equation for $\vec{L}^2$, and also

$$\frac{1}{\mathrm{i}}\frac{\partial}{\partial\phi}Y_{lm}(\theta,\phi) = mY_{lm}(\theta,\phi) \tag{5.2.37}$$

which is the eigenvalue equation for L_3.

Also worth knowing is the symmetry property

$$Y_{lm}(\theta,\phi) = Y_{l,-m}(-\theta,-\phi)\,, \tag{5.2.38}$$

which one can easily verify for $l=0,1,2$ by a quick look at

$$\begin{aligned}
l=0: \quad & Y_{00}=\sqrt{\frac{1}{4\pi}}\,; \\
l=1: \quad & Y_{10}=\sqrt{\frac{3}{4\pi}}\cos\theta=\sqrt{\frac{3}{4\pi}}\frac{x_3}{r}\,, \\
& Y_{1\pm1}=\mp\sqrt{\frac{3}{8\pi}}\sin\theta\,\mathrm{e}^{\pm\mathrm{i}\phi}=\mp\sqrt{\frac{3}{8\pi}}\frac{x_1\pm\mathrm{i}x_2}{r}\,; \\
l=2: \quad & Y_{20}=\sqrt{\frac{5}{16\pi}}\left(3\cos^2\theta-1\right)=\sqrt{\frac{5}{16\pi}}\frac{3x_3^2-r^2}{r^2}\,, \\
& Y_{2\pm1}=\mp\sqrt{\frac{15}{8\pi}}\cos\theta\sin\theta\,\mathrm{e}^{\pm\mathrm{i}\phi}=\mp\sqrt{\frac{15}{8\pi}}\frac{(x_1\pm x_2)x_3}{r^2}\,, \\
& Y_{2\pm2}=\sqrt{\frac{15}{32\pi}}\sin^2\theta\,\mathrm{e}^{\pm2\mathrm{i}\phi}=\sqrt{\frac{15}{32\pi}}\frac{\left(x_1\pm\mathrm{i}x_2\right)^2}{r^2}\,.
\end{aligned} \tag{5.2.39}$$

The versions involving the cartesian coordinates recall that

$$x_1\pm\mathrm{i}x_2=r\sin\theta\,\mathrm{e}^{\pm\mathrm{i}\phi} \quad \text{and} \quad x_3=r\cos\theta\,, \tag{5.2.40}$$

and hint at the so-called *solid harmonics*, essentially linear combinations of r^lY_{lm} with common l, that can be systematically regarded as polynomials of the form

$$(\vec{a}\cdot\vec{r})^l \quad \text{with} \quad \vec{a}\cdot\vec{a}=0\,, \tag{5.2.41}$$

where $\vec{a}$ is a *complex* vector such as $\vec{a}=\vec{e}_1+\mathrm{i}\vec{e}_2$. The basic observation in this context is

$$\vec{\nabla}^2(\vec{a}\cdot\vec{r})^l=0\,, \tag{5.2.42}$$

and this can serve as a starting point for a systematic study of the $Y_{lm}(\theta,\phi)$.

We are here content with recognizing the central statement, which we owe to John W. Strutt (Lord Rayleigh), namely the generating function

$$\frac{1}{l!}(\vec{a}\cdot\vec{r})^l=r^l\sum_{m=-l}^{l}\frac{\alpha_+^{l+m}\alpha_-^{l-m}}{\sqrt{(l+m)!(l-m)!}}\sqrt{\frac{4\pi}{2l+1}}Y_{lm}(\theta,\phi) \tag{5.2.43}$$

where α_+ and α_- are arbitrary complex numbers that parameterize the cartesian components of $\vec{a}$,

$$\vec{a}\mathrel{\widehat{=}}\left(\frac{\alpha_-^2-\alpha_+^2}{2},\frac{\alpha_-^2+\alpha_+^2}{2\mathrm{i}},\alpha_+\alpha_-\right), \tag{5.2.44}$$

and $\vec{r}$ has components

$$\vec{r} \mathrel{\widehat{=}} r(\sin\theta\cos\phi, \sin\theta\sin\phi, \cos\theta)\,, \tag{5.2.45}$$

which repeats (5.2.40) of course. All properties of the spherical harmonics — for example the orthonormality relation (5.2.35) — are contained in this generating function and can be derived from it rather easily.

Chapter 6

Approximation Methods

6.1 Hellmann–Feynman theorem

In (5.1.16) we found the *Bohr energies*

$$E_n = -\frac{Z^2}{2n^2}\frac{e^2}{a_0} \quad \text{with} \quad a_0 = \frac{\hbar^2}{Me^2} \tag{6.1.1}$$

for hydrogenic atoms, with the principal quantum number $n = n_r + l + 1 = 1, 2, 3, \ldots$ that labels the *Bohr shells.* The potential energy in the Hamilton operator (5.1.1),

$$H = \frac{1}{2M}\vec{P}^2 - \frac{Ze^2}{R}, \quad R = |\vec{R}|, \tag{6.1.2}$$

is proportional to Z whereas $E_n \propto Z^2$. This suggests that $R \propto 1/Z$, which is physically reasonable because a larger nuclear charge is expected to attract the electron more strongly and thus the size of the orbits should shrink with increasing Z.

We can make this statement more quantitative by calculating the expectation value of R with the aid of the wave functions (5.2.28), but there is a much simpler and more instructive way of going about this: We shall use the Hellmann–Feynman theorem (Hans Hellmann and Richard P. Feynman), which we first state in a general context.

Consider a Hamilton operator H_λ that depends on a parameter λ and its λ dependent eigenstates and eigenvalues,

$$H_\lambda|E_\lambda, \ldots\rangle = |E_\lambda, \ldots\rangle E_\lambda \tag{6.1.3}$$

where the ellipsis indicates other quantum numbers (such as l, m for the angular momentum) that do not depend on parameter λ. Differentiation with respect to λ establishes

$$\frac{\partial H_\lambda}{\partial \lambda}|E_\lambda,\ldots\rangle + H_\lambda \frac{\partial|E_\lambda,\ldots\rangle}{\partial \lambda} = \frac{\partial|E_\lambda,\ldots\rangle}{\partial \lambda} E_\lambda + |E_\lambda,\ldots\rangle \frac{\partial E_\lambda}{\partial \lambda} . \quad (6.1.4)$$

Now multiply by the corresponding bra from the left and get

$$\langle E_\lambda,\ldots| \frac{\partial H_\lambda}{\partial \lambda}|E_\lambda,\ldots\rangle = \frac{\partial E_\lambda}{\partial \lambda} \quad (6.1.5)$$

after taking into account that

$$\langle E_\lambda,\ldots|H = E_\lambda \langle E_\lambda,\ldots| \quad \text{and} \quad \langle E_\lambda,\ldots|E_\lambda,\ldots\rangle = 1 . \quad (6.1.6)$$

This, then, is the *Hellmann–Feynman theorem:*

$$\left\langle \frac{\partial H_\lambda}{\partial \lambda} \right\rangle = \frac{\partial E_\lambda}{\partial \lambda} \quad (6.1.7)$$

for expectation values that are taken with an eigenstate of H_λ to energy E_λ. In words: You find the derivative of E_λ with respect to parameter λ as the expectation value of the λ derivative of the Hamilton operator.

In the Hamilton operator (6.1.2) we have M and Z as parameters; they appear in the Bohr energies of (6.1.1) such that $E_n \propto Z^2 M$. Accordingly, the Hellmann–Feynman theorem tells us that

$$Z \frac{\partial}{\partial Z} E_n = 2E_n = \left\langle -\frac{Ze^2}{R} \right\rangle = E_{\text{pot}} , \quad (6.1.8)$$

and

$$M \frac{\partial}{\partial M} E_n = E_n = -\left\langle \frac{\vec{P}^2}{2M} \right\rangle = -E_{\text{kin}} . \quad (6.1.9)$$

Or

$$E_{\text{pot}} = 2E , \qquad E_{\text{kin}} = -E \quad (6.1.10)$$

for such a Coulomb system. Actually, this statement is generally true for many-particle systems when all components have Coulomb interactions between them.

We are particularly interested in the statement (6.1.8) about the potential energy because it implies

$$\left\langle \frac{1}{R} \right\rangle = \frac{Z}{n^2 a_0} , \quad (6.1.11)$$

so that, indeed, $R \cong n^2 a_0/Z \propto 1/Z$ in the nth Bohr shell. Note that this differs from what the wave function (5.2.28) seems to suggest, namely that $R \sim n a_0/Z$, linear in n rather than quadratic. One must take the detailed structure of the Laguerre polynomials into account to find the extra n factor from the wave functions. Fortunately, we do not need to do this because the Hellmann–Feynman theorem saves us the trouble.

In the harmonic oscillator,

$$H = \frac{1}{2M}P^2 + \frac{1}{2}M\omega^2 X^2,$$
$$H\big|n\big\rangle = \big|n\big\rangle\hbar\omega(n + \tfrac{1}{2}) \tag{6.1.12}$$

we have the parameters M and ω, and the energy eigenvalues $\hbar\omega(n + \frac{1}{2})$. Thus, considering the M dependence,

$$M\frac{\partial}{\partial M}E = -\left\langle \frac{1}{2M}P^2 \right\rangle + \left\langle \frac{1}{2}M\omega^2 X^2 \right\rangle \tag{6.1.13}$$

or, since $E = \hbar\omega(n + \frac{1}{2})$ does not involve M,

$$0 = -E_{\mathrm{kin}} + E_{\mathrm{pot}}\,; \tag{6.1.14}$$

and considering the ω dependence,

$$\omega\frac{\partial}{\partial \omega}E = E = 2\left\langle \frac{1}{2}M\omega^2 X^2 \right\rangle = 2E_{\mathrm{pot}}\,. \tag{6.1.15}$$

Therefore,

$$E_{\mathrm{kin}} = E_{\mathrm{pot}} = \frac{1}{2}E \tag{6.1.16}$$

applies to a harmonic oscillator. The argument here is given for a harmonic oscillator in one dimension but it extends immediately to two or more dimensions. Note that we did not subtract $\frac{1}{2}\hbar\omega$ (or $\hbar\omega$ in two dimensions, $\frac{3}{2}\hbar\omega$ in three dimension, ...) for the sake of a simple result.

We have seen here two examples of a more general result, namely

$$\begin{aligned}
&\text{Coulomb: potential energy} \propto \frac{1}{R}\,,\ E_{\mathrm{kin}} = -E\,,\ E_{\mathrm{pot}} = 2E\,,\\
&\text{Oscillator: potential energy} \propto R^2\,,\ E_{\mathrm{kin}} = \frac{1}{2}E\,,\ E_{\mathrm{pot}} = \frac{1}{2}E\,.
\end{aligned} \tag{6.1.17}$$

This invites the question: what about other powers of R?

6.2 Virial theorem

The answer is given by the *virial theorem*. We state it for Hamilton operators of the typical form

$$H = \underbrace{\frac{1}{2M}\vec{P}^2}_{\text{kinetic energy}} + \underbrace{V(\vec{R})}_{\text{potential energy}} \tag{6.2.1}$$

and later specialize to potentials $V(\vec{R})$ that are powers of $R = |\vec{R}|$.

First, recall (or note) that the expectation value of a commutator with H in an eigenstate of H always vanishes,

$$\langle E,\ldots|[A,H]|E,\ldots\rangle = \langle E,\ldots|(A\underset{\substack{\downarrow\\ E}}{H} - \underset{\substack{\downarrow\\ E}}{H}A)|E,\ldots\rangle = 0 \tag{6.2.2}$$

for *any* operator A. We apply this to

$$A = \vec{R}\cdot\vec{P}\,, \tag{6.2.3}$$

for which

$$\begin{aligned}[A,H] &= \mathrm{i}\hbar\frac{\mathrm{d}\vec{R}}{\mathrm{d}t}\cdot\vec{P} + \mathrm{i}\hbar\vec{R}\cdot\frac{\mathrm{d}\vec{P}}{\mathrm{d}t}\\ &= \mathrm{i}\hbar\frac{\vec{P}}{M}\cdot\vec{P} + \mathrm{i}\hbar\vec{R}\cdot\left(-\vec{\nabla}V(\vec{R})\right),\end{aligned} \tag{6.2.4}$$

so that

$$\left\langle\frac{1}{M}\vec{P}^2\right\rangle = \left\langle\vec{R}\cdot\vec{\nabla}V(\vec{R})\right\rangle \tag{6.2.5}$$

or

$$E_{\text{kin}} = \frac{1}{2}\left\langle\vec{R}\cdot\vec{\nabla}V(\vec{R})\right\rangle \tag{6.2.6}$$

for *any* energy eigenstate.

For potentials that are powers of R,

$$V(\vec{R}) = \kappa|\vec{R}|^n = \kappa R^n \tag{6.2.7}$$

with some strength parameter κ, we have

$$\vec{R}\cdot\vec{\nabla}V(\vec{R}) = nV(\vec{R}) \tag{6.2.8}$$

and so get

$$E_{\text{kin}} = \frac{n}{2} E_{\text{pot}} \,. \tag{6.2.9}$$

Together with $E = E_{\text{kin}} + E_{\text{pot}}$ it tells us that

$$E_{\text{kin}} = \frac{n}{n+2} E \,, \quad E_{\text{pot}} = \frac{2}{n+2} E \,. \tag{6.2.10}$$

For $n = -1$ and $n = 2$, this reproduces the results for Coulomb systems, such as hydrogenic atoms, and harmonic oscillators in (6.1.17),

$$\begin{aligned} n = -1 : &\quad E_{\text{kin}} = -E \,,\; E_{\text{pot}} = 2E \,;\; E_{\text{kin}} = -\frac{1}{2} E_{\text{pot}} \\ n = 2 : &\quad E_{\text{kin}} = \frac{1}{2} E \,,\; E_{\text{pot}} = \frac{1}{2} E \,;\; E_{\text{kin}} = E_{\text{pot}} \,. \end{aligned} \tag{6.2.11}$$

In another way of looking at this, we consider small changes of the kets and bras with which the expectation values are taken,

$$\big| \;\big\rangle \to \mathrm{e}^{\mathrm{i}\epsilon G} \big| \;\big\rangle \,, \quad \big\langle \;\big| \to \big\langle \;\big| \mathrm{e}^{-\mathrm{i}\epsilon G} \,, \tag{6.2.12}$$

where $G = G^\dagger$ is the hermitian generator of the unitary transformation, and ϵ is a small parameter. We keep terms up to ϵ^2 only,

$$\begin{aligned} \mathrm{e}^{\mathrm{i}\epsilon G} &= 1 + \mathrm{i}\epsilon G - \frac{1}{2}\epsilon^2 G^2 \,, \\ \mathrm{e}^{-\mathrm{i}\epsilon G} &= 1 - \mathrm{i}\epsilon G - \frac{1}{2}\epsilon^2 G^2 \,, \end{aligned} \tag{6.2.13}$$

as they will be all we need in the present context. The ϵ-dependent expectation value of H is then

$$\begin{aligned} \langle H \rangle_\epsilon &= \left\langle \mathrm{e}^{-\mathrm{i}\epsilon G} H \,\mathrm{e}^{\mathrm{i}\epsilon G} \right\rangle \\ &= \left\langle \left(1 - \mathrm{i}\epsilon G - \frac{1}{2}\epsilon^2 G^2\right) H \left(1 + \mathrm{i}\epsilon G - \frac{1}{2}\epsilon^2 G^2\right) \right\rangle \\ &= \langle H \rangle + \epsilon \left\langle \mathrm{i}[H, G] \right\rangle \\ &\quad - \frac{1}{2}\epsilon^2 \Big\langle \left(G^2 H - 2GHG + HG^2\right) \Big\rangle \,. \end{aligned} \tag{6.2.14}$$

We take the expectation value in an eigenstate of H to eigenvalue E, $\big| \;\big\rangle = \big| E, \ldots \big\rangle$, so that

$$\langle H \rangle_\epsilon = E + \frac{1}{2}\epsilon^2 \Big\langle \left(2GHG - HG^2 - G^2 H\right) \Big\rangle + O(\epsilon^3) \,, \tag{6.2.15}$$

where the linear term, $\propto \epsilon$, is absent since

$$\langle E,\ldots\big|\mathrm{i}[H,G]\big|E,\ldots\rangle = 0\,, \tag{6.2.16}$$

which repeats (6.2.2) for $A = G$. The conclusion is that

$$\frac{\partial}{\partial\epsilon}\,\langle H\rangle_{\epsilon}\bigg|_{\epsilon\,=\,0} = 0\,. \tag{6.2.17}$$

We exploit this now for the unitary *scaling transformation*

$$\begin{aligned}\vec{P} &\to \mathrm{e}^{\epsilon}\vec{P} = \mathrm{e}^{-\mathrm{i}\epsilon G}\vec{P}\,\mathrm{e}^{\mathrm{i}\epsilon G}\,,\\ \vec{R} &\to \mathrm{e}^{-\epsilon}\vec{R} = \mathrm{e}^{-\mathrm{i}\epsilon G}\vec{R}\,\mathrm{e}^{\mathrm{i}\epsilon G}\,,\end{aligned} \tag{6.2.18}$$

(see Exercise 4-1 on page 109) with

$$G = \frac{1}{2\hbar}\big(\vec{R}\cdot\vec{P} + \vec{P}\cdot\vec{R}\big) = \frac{1}{\hbar}\vec{R}\cdot\vec{P} - \frac{3}{2}\mathrm{i}\,, \tag{6.2.19}$$

so that

$$\mathrm{e}^{-\mathrm{i}\epsilon G}H\,\mathrm{e}^{\mathrm{i}\epsilon G} = \frac{1}{2M}\big(\mathrm{e}^{\epsilon}\vec{P}\big)^2 + V\big(\mathrm{e}^{-\epsilon}\vec{R}\big) \tag{6.2.20}$$

and

$$\langle H\rangle_{\epsilon} = \mathrm{e}^{2\epsilon}\Big\langle\frac{1}{2M}\vec{P}^2\Big\rangle + \Big\langle V\big(\mathrm{e}^{-\epsilon}\vec{R}\big)\Big\rangle\,. \tag{6.2.21}$$

Now differentiate with respect to ϵ,

$$\frac{\partial}{\partial\epsilon}\,\langle H\rangle_{\epsilon}\bigg|_{\epsilon=0} = 2E_{\mathrm{kin}} - \Big\langle\vec{R}\cdot\vec{\nabla}V(\vec{R})\Big\rangle = 0\,, \tag{6.2.22}$$

and arrive again at the statement of the virial theorem

$$E_{\mathrm{kin}} = \frac{1}{2}\Big\langle\vec{R}\cdot\vec{\nabla}V(\vec{R})\Big\rangle = -\frac{1}{2}\Big\langle\vec{R}\cdot\vec{F}(\vec{R})\Big\rangle\,, \tag{6.2.23}$$

the latter version using the force operator

$$\vec{F}(\vec{R}) = -\vec{\nabla}V(\vec{R}) = \frac{1}{\mathrm{i}\hbar}\Big[\vec{P},V(\vec{R})\Big]\,. \tag{6.2.24}$$

What can we say about the term $\propto \epsilon^2$ in (6.2.15)? We write

$$\begin{aligned}&2GHG - HG^2 - G^2H\\ &\quad= 2G(H-E)G - (H-E)G^2 - G^2(H-E)\,,\end{aligned} \tag{6.2.25}$$

which is a simple but useful identity because if we now take the expectation value $\langle E,\ldots|\cdots|E,\ldots\rangle$ in an eigenstate of H the last two terms vanish, and so we have

$$\frac{1}{2}\epsilon^2\Big\langle\big(2GHG - HG^2 - G^2H\big)\Big\rangle = \epsilon^2\langle E,\gamma|G(H-E)G|E,\gamma\rangle \qquad (6.2.26)$$

where the symbol γ stands for all the other quantum numbers that we might need to specify in the case of an energetic degeneracy. Using the completeness of the eigenstates of H,

$$H - E = \sum_{E',\gamma'} |E',\gamma'\rangle(E'-E)\langle E',\gamma'|\,, \qquad (6.2.27)$$

we further have

$$\begin{aligned}&\frac{1}{2}\epsilon^2\Big\langle\big(2GHG - HG^2 - G^2H\big)\Big\rangle\\ &\qquad = \epsilon^2\sum_{E',\gamma'}(E'-E)\underbrace{\big|\langle E,\gamma|G|E',\gamma'\rangle\big|^2}_{\geq 0}\,. \end{aligned} \qquad (6.2.28)$$

The summation is over all eigenvalues E' of H (and, if necessary, over the γ' quantum numbers), whereby the terms with $E' = E$ do not contribute. This implies that the ϵ^2 term is nonnegative if E is the smallest eigenvalue of H, that is if E is the ground-state energy.

The conclusion is that $\langle \mathrm{e}^{-\mathrm{i}\epsilon G}H\,\mathrm{e}^{\mathrm{i}\epsilon G}\rangle$ has a minimum at $\epsilon = 0$ if the expectation value is taken in the ground state of H. For all other eigenstates of H, we have an extremum at $\epsilon = 0$, but it could be either a minimum or a maximum, depending on the value of the sum over E' and γ'.

6.3 Rayleigh–Ritz variational method

Any ket $|\;\rangle$ can be related to the ket $|E_0\rangle$ of the (nondegenerate) ground state,

$$|\;\rangle = U|E_0\rangle\,, \qquad (6.3.1)$$

by a suitable unitary tranformation U (which, incidentally, is not unique). It follows that

$$\langle H\rangle \geq E_0 \qquad (6.3.2)$$

for *any* ket $|\;\rangle$. This is known as the *Rayleigh–Ritz variational method*, named after Lord Rayleigh and Walther Ritz. It is more generally true than the above derivation suggests, because the limitation to ϵ^2 terms is not necessary. We just need to use the completeness of the $|E,\gamma\rangle$ states in

$$\langle H\rangle = \sum_{E,\gamma} E\big|\langle E,\gamma|\ \rangle\big|^2 \tag{6.3.3}$$

and the fact that $E \geq E_0$ for all E,

$$\langle H\rangle \geq \sum_{E,\gamma} E_0\big|\langle E,\gamma|\ \rangle\big|^2 = E_0\,. \tag{6.3.4}$$

This offers a very convenient method for obtaining very good estimates of ground-state energies without actually solving the Schrödinger eigenvalue equation.

As an example let us consider a one-dimensional system with a constant restoring force, for which the Hamilton operator is

$$H = \frac{1}{2M}P^2 + F|X|\,. \tag{6.3.5}$$

A trial ket $|\ \rangle$ is specified by its *trial wave function* in position,

$$\psi(x) = \langle x|\ \rangle \tag{6.3.6}$$

which is normalized, $\langle\ |\ \rangle = 1$ or

$$\int \mathrm{d}x\, |\psi(x)|^2 = 1\,. \tag{6.3.7}$$

The Rayleigh–Ritz method then exploits

$$\int \mathrm{d}x \left(\frac{\hbar^2}{2M}\left|\frac{\partial\psi(x)}{\partial x}\right|^2 + F|x|\,|\psi(x)|^2\right) \geq E_0 \tag{6.3.8}$$

where E_0 is the ground-state energy of the Hamilton operator that we are considering. The choice of trial function is usually dictated by (i) our intuition about the shape of the ground-state wave function and (ii) practical considerations, after all we want to be able to execute the required integrations. Therefore, the trial wave function should be reasonably simple.

So, why not try

$$\psi(x) = \sqrt{\kappa}\,\mathrm{e}^{-\kappa|x|}\,, \quad \kappa > 0\,, \tag{6.3.9}$$

with an adjustable parameter κ? This gives

$$\begin{aligned}
E_{\text{kin}} &= \left\langle \frac{1}{2M} P^2 \right\rangle = \frac{\hbar^2}{2M} \int \mathrm{d}x \left| \frac{\partial \psi(x)}{\partial x} \right|^2 \\
&= \frac{\hbar^2}{2M} \int \mathrm{d}x \left| -\sqrt{\kappa^3}\, \mathrm{sgn}\,(x)\, \mathrm{e}^{-\kappa|x|} \right|^2 \\
&= \frac{\hbar^2}{2M} \kappa^2 \underbrace{\int \mathrm{d}x\, \kappa\, \mathrm{e}^{-2\kappa|x|}}_{= \int \mathrm{d}x\, |\psi(x)|^2 = 1} \\
&= \frac{(\hbar\kappa)^2}{2M}
\end{aligned} \tag{6.3.10}$$

for the kinetic energy, and

$$\begin{aligned}
E_{\text{pot}} &= \Big\langle F|X| \Big\rangle = F \int \mathrm{d}x\, |x|\, |\psi(x)|^2 \\
&= F\kappa \int \mathrm{d}x\, |x|\, \mathrm{e}^{-2\kappa|x|} \\
&= F\kappa \left(-\frac{1}{2} \frac{\partial}{\partial \kappa} \right) \underbrace{\int \mathrm{d}x\, \mathrm{e}^{-2\kappa|x|}}_{= \kappa^{-1} \int \mathrm{d}x\, |\psi(x)|^2 = \kappa^{-1}} \\
&= F\kappa \left(-\frac{1}{2} \frac{\partial}{\partial \kappa} \right) \frac{1}{\kappa} = \frac{F}{2\kappa}
\end{aligned} \tag{6.3.11}$$

for the potential energy. Taken together, we have

$$\langle H \rangle = E_{\text{kin}} + E_{\text{pot}} = \frac{(\hbar\kappa)^2}{2M} + \frac{F}{2\kappa} \geq E_0 \tag{6.3.12}$$

and this inequality must be true irrespective of the value we choose for κ on the left.

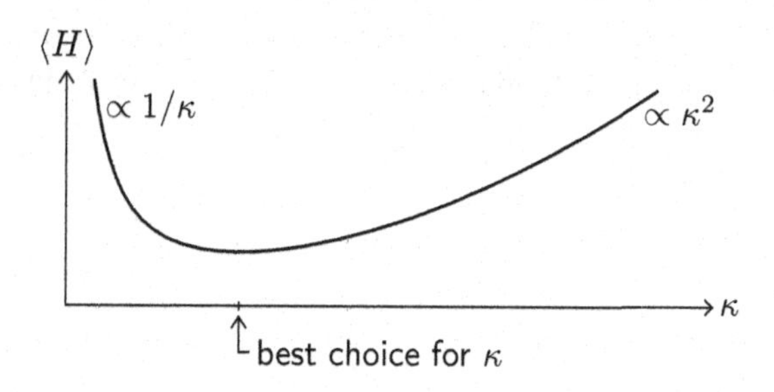

The best choice for κ is thus the value for which the left-hand side gives the *lowest upper bound* on the ground-state energy E_0. We differentiate with respect to κ,

$$\frac{\partial}{\partial\kappa}\langle H\rangle = \frac{\hbar^2}{M}\kappa - \frac{F}{2\kappa^2}\,, \tag{6.3.13}$$

which has to vanish for the best choice for κ. Accordingly,

$$\kappa = \left(\frac{MF}{2\hbar^2}\right)^{1/3} \tag{6.3.14}$$

is the optimal κ value. For this κ, then,

$$E_{\text{kin}} = \frac{\hbar^2}{2M}\left(\frac{MF}{2\hbar^2}\right)^{2/3} = \left(\frac{1}{2}\right)^{5/3}\left(\frac{\hbar^2F^2}{M}\right)^{1/3} \tag{6.3.15}$$

and

$$E_{\text{pot}} = \frac{1}{2}F\left(\frac{2\hbar^2}{MF}\right)^{1/3} = \left(\frac{1}{2}\right)^{2/3}\left(\frac{\hbar^2F^2}{M}\right)^{1/3}, \tag{6.3.16}$$

giving

$$E_0 \leq \langle H\rangle = \frac{3}{2^{5/3}}\left(\frac{\hbar^2F^2}{M}\right)^{1/3} = 0.945\left(\frac{\hbar^2F^2}{M}\right)^{1/3}. \tag{6.3.17}$$

The upper bound thus found exceeds the actual ground-state energy

$$E_0 = 0.80862\left(\frac{\hbar^2F^2}{M}\right)^{1/3} \tag{6.3.18}$$

by almost 17%, which is a very poor performance for a variational estimate. So, to not lose face, we do not tell anybody and try to do better.

Part of the procedure is the choice of the best scale, the choice of the optimal value for κ. Let us be more systematic about this from the outset by considering not one particular trial function, but a whole family of them,

$$\langle x|\ \rangle = C\psi(\kappa x) \tag{6.3.19}$$

with the normalizing prefactor such that

$$1 = \langle\ |\ \rangle = \int \mathrm{d}x\,|C\psi(\kappa x)|^2 = \frac{1}{\kappa}|C|^2\int \mathrm{d}y\,|\psi(y)|^2 \tag{6.3.20}$$

(substitute $\kappa x = y$, $\kappa \mathrm{d}x = \mathrm{d}y$) or

$$\frac{\kappa}{|C|^2} = \int \mathrm{d}y\, |\psi(y)|^2 . \tag{6.3.21}$$

For the kinetic energy we now have

$$\begin{aligned} E_{\text{kin}} &= \frac{\hbar^2}{2M}|C|^2 \int \mathrm{d}x \left|\frac{\partial}{\partial x}\psi(\kappa x)\right|^2 = \frac{\hbar^2}{2M}|C|^2\kappa \int \mathrm{d}y \left|\frac{\mathrm{d}\psi(y)}{\mathrm{d}y}\right|^2 \\ &= \frac{(\hbar\kappa)^2}{2M}\frac{\int \mathrm{d}y \left|\frac{\mathrm{d}\psi(y)}{\mathrm{d}y}\right|^2}{\int \mathrm{d}y\, |\psi(y)|^2}, \end{aligned} \tag{6.3.22}$$

and we get

$$\begin{aligned} E_{\text{pot}} &= F|C|^2 \int \mathrm{d}x\, |x|\, |\psi(\kappa x)|^2 = F|C|^2 \frac{1}{\kappa^2} \int \mathrm{d}y\, |y| |\psi(y)|^2 \\ &= \frac{F}{\kappa}\frac{\int \mathrm{d}y\, |y|\, |\psi(y)|^2}{\int \mathrm{d}y\, |\psi(y)|^2} \end{aligned} \tag{6.3.23}$$

for the potential energy. Together they give us the upper bound on E_0,

$$\begin{aligned} E_0 \le \langle H \rangle &= E_{\text{kin}} + E_{\text{pot}} \\ &= \frac{(\hbar\kappa)^2}{2M} \underbrace{\frac{\int \mathrm{d}y \left|\frac{\mathrm{d}\psi(y)}{\mathrm{d}y}\right|^2}{\int \mathrm{d}y |\psi(y)|^2}}_{\equiv \boxed{1}} + \frac{F}{\kappa} \underbrace{\frac{\int \mathrm{d}y\, |y|\, |\psi(y)|^2}{\int \mathrm{d}y\, |\psi(y)|^2}}_{\equiv \boxed{2}} \\ &= \frac{(\hbar\kappa)^2}{2M}\boxed{1} + \frac{F}{\kappa}\boxed{2} \end{aligned} \tag{6.3.24}$$

and we find the best choice for κ by differentiation as we did above. So, for the optimal κ value

$$\frac{\hbar^2\kappa}{M}\boxed{1} - \frac{F}{\kappa^2}\boxed{2} = 0 \tag{6.3.25}$$

with the consequence

$$\kappa = \left(\frac{MF}{\hbar^2}\right)^{1/3} \left(\frac{\boxed{2}}{\boxed{1}}\right)^{1/3}, \tag{6.3.26}$$

so that

$$E_{\text{kin}} = \frac{1}{2}\left(\frac{\hbar^2 F^2}{M}\right)^{1/3} \boxed{1}^{1/3} \boxed{2}^{2/3} \qquad (6.3.27)$$

and

$$E_{\text{pot}} = \left(\frac{\hbar^2 F^2}{M}\right)^{1/3} \boxed{1}^{1/3} \boxed{2}^{2/3} \qquad (6.3.28)$$

for this optimally chosen κ.

Taking everything together, we have

$$E_0 \le \frac{3}{2}\left(\frac{\hbar^2 F^2}{M}\right)^{1/3} \boxed{1}^{1/3} \boxed{2}^{2/3} \qquad (6.3.29)$$

or

$$\frac{E_0}{(\hbar^2 F^2/M)^{1/3}} \le \frac{3}{2} \frac{\left[\displaystyle\int \mathrm{d}y \left|\frac{\mathrm{d}\psi(y)}{\mathrm{d}y}\right|^2\right]^{1/3} \left[\displaystyle\int \mathrm{d}y\, |y|\, |\psi(y)|^2\right]^{2/3}}{\displaystyle\int \mathrm{d}y |\psi(y)|^2} . \qquad (6.3.30)$$

In this form, both the normalization of $C\psi(\kappa x)$ to unit integral of its squared modulus and the optimal choice of the scale are built into the functional on the right, which is thus *scale invariant.* This is to say that the right-hand side gives the same values for, say,

$$\psi(y) = 2\,\mathrm{e}^{-4|y|} \quad \text{and} \quad \psi(y) = \mathrm{e}^{-3|y|} , \qquad (6.3.31)$$

the first of which is normalized to $\int \mathrm{d}y |\psi(y)|^2 = 1$, the other not. Accordingly, we need not bother anymore about the normalization and the scale of the trial wave function, both have been taken care of once and for all.

For either one of the wave functions in (6.3.31) we get, of course, the rather poor estimate of (6.3.17). Let us see how we fare with a gaussian trial function instead,

$$\begin{aligned} \psi(y) &= \mathrm{e}^{-\frac{1}{2}y^2} , \\ \frac{\mathrm{d}}{\mathrm{d}y}\psi(y) &= -y\,\mathrm{e}^{-\frac{1}{2}y^2} , \end{aligned} \qquad (6.3.32)$$

where the $\frac{1}{2}$ is for convenience (we exploit the scale invariance here). For this $\psi(y)$, we have

$$\int \mathrm{d}y\,|\psi(y)|^2 = \int \mathrm{d}y\,\mathrm{e}^{-y^2} = \sqrt{\pi}\,,$$
$$\int \mathrm{d}y \left|\frac{\mathrm{d}\psi(y)}{\mathrm{d}y}\right|^2 = \int \mathrm{d}y\, y^2\,\mathrm{e}^{-y^2} = \frac{1}{2}\sqrt{\pi} \tag{6.3.33}$$

(use $y^2\,\mathrm{e}^{-y^2} = -\frac{1}{2}y\frac{\mathrm{d}}{\mathrm{d}y}\,\mathrm{e}^{-y^2}$ and integrate by parts) and

$$\int \mathrm{d}y\,|y|\,|\psi(y)|^2 = 2\int_0^\infty \mathrm{d}y\,\underbrace{y\,\mathrm{e}^{-y^2}}_{=-\frac{1}{2}\frac{\mathrm{d}}{\mathrm{d}y}\,\mathrm{e}^{-y^2}} = 1\,. \tag{6.3.34}$$

The resulting estimate for E_0,

$$\frac{E_0}{\left(\frac{\hbar^2F^2}{M}\right)^{1/3}} \leq \frac{3}{2}\frac{\left(\frac{1}{2}\sqrt{\pi}\right)^{1/3}1^{2/3}}{\sqrt{\pi}} = \frac{3}{2^{4/3}}\pi^{-1/3}$$
$$= 0.8129 = 1.0053 \times 0.80862 \tag{6.3.35}$$

is only 0.53% in excess of the exact value.

6-1 Try $\psi(y) = \mathrm{e}^{-|y|^\gamma}$ with different values for γ. Use Leonhard Euler's factorial integral

$$\int_0^\infty \mathrm{d}x\,x^\nu\,\mathrm{e}^{-x} = \nu!\,, \qquad \nu > -1\,,$$

to express the upper bound in terms of factorials (of noninteger numbers). For $\gamma = 1$ and $\gamma = 2$ this reproduces the values of (6.3.17) and (6.3.35), respectively. But what do you get for $\gamma = \frac{3}{2}$ or $\gamma = \frac{7}{4}$?

6-2 Find the scale-invariant expression for the upper bound for

$$H = \frac{1}{2M}P^2 + \frac{\lambda}{n}|X|^n\,, \qquad \lambda > 0\,,$$

with arbitrary power $n > 0$. Try the trial wave functions $\psi(y) = \mathrm{e}^{-|y|}$ and $\psi(y) = \mathrm{e}^{-\frac{1}{2}y^2}$ and compare the upper bounds for $n = 2, 3$, and 4.

6-3 Find an upper bound on the ground-state energy of

$$H = \frac{1}{2M}P^2 - \frac{(\hbar\kappa)^2}{2M}\,\mathrm{e}^{-2\kappa|X|}$$

with $\kappa > 0$; try to do well with a "simple" trial function.

6-4 Use gaussian trial wave functions to find an upper bound on the ground-state energy of

$$H = \frac{1}{2M}P^2 - \frac{(\hbar\kappa)^2}{\sqrt{2}M}\,\mathrm{e}^{-\frac{1}{2}(\kappa X)^2}$$

with $\kappa > 0$.

Another application, to two-electron atoms, is found in Section 6.4.1 of *Perturbed Evolution*.

One can also apply the Rayleigh–Ritz method to excited states, but then one must restrict the trial wave functions. Return to (6.3.3) and note that

$$\begin{aligned}\langle H\rangle &= \sum_{E,\gamma} E\left|\langle E,\gamma|\;\rangle\right|^2 \\ &\geq \sum_{E,\gamma} E_1\left|\langle E,\gamma|\;\rangle\right|^2 = E_1 \end{aligned} \tag{6.3.36}$$

if $\left|\langle E_0,\gamma|\;\rangle\right|^2 = 0$ so that the term with $E = E_0$ does not contribute. In view of

$$E_0 < E_1 < E_2 < \cdots \tag{6.3.37}$$

(ground state: E_0; 1st excited state: E_1; 2nd excited state: E_2)

all other terms then have $E \geq E_1$ and

$$\langle H\rangle \geq E_1 \quad \text{for} \quad \langle E_0,\gamma|\;\rangle = 0 \tag{6.3.38}$$

follows.

This is, however, of possibly limited use because we do not know the ground-state ket $|E_0,\gamma\rangle$ precisely. But such precise knowledge may not be necessary if we can exploit certain known properties of the ground state to ensure that all trial kets considered are orthogonal to the ground-state ket.

Take the example of the constant restoring force again. The Hamilton operator (6.3.5) has a reflection symmetry, which is to say that it is invariant under the unitary transformation

$$X \to -X\,, \quad P \to -P\,. \tag{6.3.39}$$

As a consequence, the wave functions of its eigenstates are either even or odd,

$$\text{even wave function: } \langle -x|E\rangle = \langle x|E\rangle\,,$$
$$\text{odd wave function: } \langle -x|E\rangle = -\langle x|E\rangle\,, \qquad (6.3.40)$$

and the ground-state wave function will be even. By contrast, the first excited state has an odd wave function, the second excited state is even, and so forth,

$$\langle -x|E_n\rangle = (-1)^n\langle x|E_n\rangle\,. \qquad (6.3.41)$$

Now, all odd wave functions are orthogonal to all even wave functions and, therefore, we can enforce $\langle E_0|\ \rangle = 0$ here by restricting $\langle x|\ \rangle$ to odd wave functions. So, if we only allow odd trial functions,

$$\psi(y) = -\psi(-y) \qquad (6.3.42)$$

on the right-hand side of inequality (6.3.30), we get an upper bound on $\left(\hbar^2F^2/M\right)^{-1/3}E_1$.

6-5 Use $\psi(y) = y\,\mathrm{e}^{-\frac{1}{2}y^2}$ to estimate E_1 in this way, and compare with the exact answer

$$E_1 = 1.85576\left(\frac{\hbar^2F^2}{M}\right)^{1/3}.$$

6.4 Rayleigh–Schrödinger perturbation theory

The examples we have met so far, including the constant restoring force of (6.3.5), are of the kind that the eigenkets and eigenvalues of the Hamilton operator can be found exactly. One must appreciate, however, that this situation is an exception, not the rule. Much more typical in applications is the case that we cannot solve the eigenvalue problem exactly, and then one must resort to methods of approximation. The Rayleigh–Ritz variational method is of this kind, but there are also others, which can give, with sufficient effort, approximate solutions of any desired accuracy.

As a first example of such a method we consider the *Rayleigh–Schrödinger perturbation theory*, named after Lord Rayleigh and Erwin Schrödinger. It applies, in its simplest version, to Hamilton operators that can be split into a "big" part H_0, and a "small" part H_1,

$$H = H_0 + H_1 \tag{6.4.1}$$

where the "big" part H_0 has exactly known eigenvalues and eigenkets

$$H_0 \big| n^{(0)} \big\rangle = \big| n^{(0)} \big\rangle E_n^{(0)} \,, \qquad n = 0, 1, 2, \ldots \tag{6.4.2}$$

with *nondegenerate* energies,

$$E_0^{(0)} < E_1^{(0)} < E_2^{(0)} < \cdots \,, \tag{6.4.3}$$

and the "small" part H_1 is difficult to handle exactly. We denote the eigenkets of the full Hamilton operator H by $\big| n \big\rangle$ and the eigenvalues by E_n,

$$H \big| n \big\rangle = \big| n \big\rangle E_n \,, \tag{6.4.4}$$

and for later convenience, we normalize the kets $\big| n \big\rangle$ such that

$$\big\langle n^{(0)} \big| n \big\rangle = 1 \,. \tag{6.4.5}$$

By contrast, the unperturbed eigenkets of H_0 are orthonormal as usual,

$$\big\langle n^{(0)} \big| m^{(0)} \big\rangle = \delta_{nm} \,. \tag{6.4.6}$$

We now introduce a hierarchy of approximations by writing

$$\begin{aligned} E_n &= \underset{\text{0th}}{E_n^{(0)}} + \underset{\text{1st}}{E_n^{(1)}} + \underset{\text{2nd}}{E_n^{(2)}} + \cdots \quad \cdots \text{ order in } H_1 \\ \big| n \big\rangle &= \big| n^{(0)} \big\rangle + \big| n^{(1)} \big\rangle + \big| n^{(2)} \big\rangle + \cdots \end{aligned} \tag{6.4.7}$$

and insert this into the eigenvalue equation (6.4.4) to arrive at

$$\begin{aligned} &(H_0 + H_1)\Big(\big| n^{(0)} \big\rangle + \big| n^{(1)} \big\rangle + \big| n^{(2)} \big\rangle + \cdots\Big) \\ &= \Big(\big| n^{(0)} \big\rangle + \big| n^{(1)} \big\rangle + \big| n^{(2)} \big\rangle + \cdots\Big)\Big(E_n^{(0)} + E_n^{(1)} + E_n^{(2)} + \cdots\Big) \,. \end{aligned} \tag{6.4.8}$$

The 0th-order terms give

$$H_0 \big| n^{(0)} \big\rangle = \big| n^{(0)} \big\rangle E_n^{(0)} \,, \tag{6.4.9}$$

which is nothing new. Now, take the 1st-order terms and establish

$$H_0 \big| n^{(1)} \big\rangle + H_1 \big| n^{(0)} \big\rangle = \big| n^{(1)} \big\rangle E_n^{(0)} + \big| n^{(0)} \big\rangle E_n^{(1)} \,. \tag{6.4.10}$$

In view of the normalization condition (6.4.5) we have

$$1 = \langle n^{(0)} | n \rangle = \underbrace{\langle n^{(0)} | n^{(0)} \rangle}_{=1} + \underbrace{\langle n^{(0)} | n^{(1)} \rangle + \langle n^{(0)} | n^{(2)} \rangle + \dots}_{=0} \qquad (6.4.11)$$

where the $\langle n^{(0)} | n^{(k)} \rangle$ terms must vanish order by order for $k > 0$, that is to say that

$$\langle n^{(0)} | n^{(1)} \rangle = 0 \,, \quad \langle n^{(0)} | n^{(2)} \rangle = 0 \,, \quad \dots \,. \qquad (6.4.12)$$

We multiply the 1st-order ket equation (6.4.10) by bra $\langle n^{(0)} |$ and get

$$\underbrace{\langle n^{(0)} | H_0 | n^{(1)} \rangle}_{= E_n^{(0)} \langle n^{(0)} | n^{(1)} \rangle = 0} + \langle n^{(0)} | H_1 | n^{(0)} \rangle = \underbrace{\langle n^{(0)} | n^{(1)} \rangle}_{=0} E_n^{(0)} + \underbrace{\langle n^{(0)} | n^{(0)} \rangle}_{=1} E_n^{(1)} \,, \qquad (6.4.13)$$

that is

$$E_n^{(1)} = \langle n^{(0)} | H_1 | n^{(0)} \rangle \,, \qquad (6.4.14)$$

which is, in fact, the statement of the Hellmann–Feynman theorem.

6-6 Justify this remark be deriving (6.4.14) as a consequence of the Hellmann–Feynman theorem.

Then, upon multiplying by bra $\langle m^{(0)} |$ with $m \neq n$,

$$\underbrace{\langle m^{(0)} | H_0 | n^{(1)} \rangle}_{= E_m^{(0)} \langle m^{(0)} |} + \langle m^{(0)} | H_1 | n^{(0)} \rangle = \langle m^{(0)} | n^{(1)} \rangle E_n^{(0)} \,, \qquad (6.4.15)$$

we obtain

$$\langle m^{(0)} | n^{(1)} \rangle = - \frac{\langle m^{(0)} | H_1 | n^{(0)} \rangle}{E_m^{(0)} - E_n^{(0)}} \,. \qquad (6.4.16)$$

We combine this with

$$\begin{aligned} | n^{(1)} \rangle &= \sum_m | m^{(0)} \rangle \underbrace{\langle m^{(0)} | n^{(1)} \rangle}_{=0 \text{ for } m = n} \\ &= \sum_{m (\neq n)} | m^{(0)} \rangle \langle m^{(0)} | n^{(1)} \rangle \end{aligned} \qquad (6.4.17)$$

to get the 1st-order correction of the ket,

$$\left|n^{(1)}\right\rangle = -\sum_{m(\neq n)} \left|m^{(0)}\right\rangle \frac{\left\langle m^{(0)}\middle|H_1\middle|n^{(0)}\right\rangle}{E_m^{(0)} - E_n^{(0)}} . \tag{6.4.18}$$

Having found the expressions for the 1st-order corrections, we turn to the 2nd-order terms in (6.4.8),

$$H_0\left|n^{(2)}\right\rangle + H_1\left|n^{(1)}\right\rangle = \left|n^{(2)}\right\rangle E_n^{(0)} + \left|n^{(1)}\right\rangle E_n^{(1)} + \left|n^{(0)}\right\rangle E_n^{(2)} . \tag{6.4.19}$$

Multiply by bra $\left\langle n^{(0)}\right|$ to extract $E_n^{(2)}$,

$$\left\langle n^{(0)}\middle|H_1\middle|n^{(1)}\right\rangle = E_n^{(2)} , \tag{6.4.20}$$

where we need the 1st-order correction $\left|n^{(1)}\right\rangle$ of (6.4.18). Thus

$$E_n^{(2)} = -\sum_{m(\neq n)} \frac{\left|\left\langle m^{(0)}\middle|H_1\middle|n^{(0)}\right\rangle\right|^2}{E_m^{(0)} - E_n^{(0)}} . \tag{6.4.21}$$

Here, and in the earlier result (6.4.18) for $\left|n^{(1)}\right\rangle$, it is important that all $E_n^{(0)}$ energies are different, because otherwise we are bound to divide by zero eventually.

If $n = 0$, then $E_m^{(0)} - E_n^{(0)} = E_m^{(0)} - E_0^{(0)} > 0$ for all $m \neq 0$, so that

$$E_0^{(2)} = -\sum_{m>0} \frac{\left|\left\langle m^{(0)}\middle|H_1\middle|0^{(0)}\right\rangle\right|^2}{E_m^{(0)} - E_0^{(0)}} \leq 0 . \tag{6.4.22}$$

This says that the 2nd-order correction to the ground-state energy is always negative.

We can understand this fact as a consequence of a Rayleigh–Ritz estimation. Recall that

$$\langle H_0 + H_1\rangle \geq E_0 \tag{6.4.23}$$

for *any* trial ket, in particular then for $\left|0^{(0)}\right\rangle$, the unperturbed ground state, for which

$$\begin{aligned}\langle H_0 + H_1\rangle &= \left\langle 0^{(0)}\middle|H_0\middle|0^{(0)}\right\rangle + \left\langle 0^{(0)}\middle|H_1\middle|0^{(0)}\right\rangle \\ &= E_0^{(0)} + E_0^{(1)} .\end{aligned} \tag{6.4.24}$$

So,

$$E_0^{(0)} + E_0^{(1)} \geq E_0 = E_0^{(0)} + E_0^{(1)} + E_0^{(2)} + \cdots \tag{6.4.25}$$

which implies that $E_0^{(2)} + \cdots \leq 0$. But here $E_0^{(2)}$ is the dominating term and, therefore, must be negative itself: $E_0^{(2)} \leq 0$, indeed.

6-7 Find an explicit expression for $|n^{(2)}\rangle$.

6-8 We found

$$E_n^{(1)} = \langle n^{(0)}|H_1|n^{(0)}\rangle \quad \text{and} \quad E_n^{(2)} = \langle n^{(0)}|H_1|n^{(1)}\rangle ;$$

show that, more generally,

$$E_n^{(k)} = \langle n^{(0)}|H_1|n^{(k-1)}\rangle$$

holds.

Here is a simple illustrating example. Consider a harmonic oscillator with a small anharmonic perturbation,

$$\begin{aligned} H &= \frac{1}{2M}P^2 + \frac{1}{2}M\omega^2 X^2 - \frac{1}{2}\hbar\omega + \lambda\hbar\omega\left(\frac{X}{l}\right)^4 \\ &= \underbrace{\hbar\omega A^\dagger A}_{=H_0} + \underbrace{\frac{1}{4}\lambda\hbar\omega\left(A^\dagger + A\right)^4}_{=H_1} \end{aligned} \tag{6.4.26}$$

where we measure the strength of the perturbation by the (small) dimensionless parameter λ. The unperturbed states are the eigenstates of $H_0 = \hbar\omega A^\dagger A$, so they are the familiar Fock states,

$$|n^{(0)}\rangle = \frac{1}{\sqrt{n!}} A^{\dagger n} \underbrace{|0^{(0)}\rangle}_{\text{oscillator ground state}} \tag{6.4.27}$$

and

$$H_1 = \frac{1}{4}\lambda\hbar\omega\left(A^\dagger + A\right)^4 = \frac{1}{4}\lambda\hbar\omega\left(A^\dagger + A\right)^2\left(A^\dagger + A\right)^2 . \tag{6.4.28}$$

Since $A^\dagger$, A are the ladder operators for the unperturbed states $|n^{(0)}\rangle$, we have

$$\begin{aligned} \left(A^\dagger + A\right)^2|n^{(0)}\rangle &= \left({A^\dagger}^2 + A^\dagger A + AA^\dagger + A^2\right)|n^{(0)}\rangle \\ &= |(n+2)^{(0)}\rangle\sqrt{(n+1)(n+2)} + |n^{(0)}\rangle(2n+1) \\ &\quad + |(n-2)^{(0)}\rangle\sqrt{n(n-1)} , \end{aligned} \tag{6.4.29}$$

and this now gives us

$$
\begin{aligned}
E_n^{(1)} &= \langle n^{(0)}|H_1|n^{(0)}\rangle \\
&= \frac{1}{4}\lambda\hbar\omega\,\langle n^{(0)}|(A^\dagger + A)^2(A^\dagger + A)^2|n^{(0)}\rangle \\
&= \frac{1}{4}\lambda\hbar\omega\Big[(n+1)(n+2) + (2n+1)^2 + n(n-1)\Big] \\
&= \frac{3}{4}\lambda\hbar\omega\big[n^2 + (n+1)^2\big]\,, \qquad (6.4.30)
\end{aligned}
$$

so that

$$
E_n = \hbar\omega\left(n + \frac{3}{4}\lambda\big[n^2 + (n+1)^2\big] + O(\lambda^2)\right). \qquad (6.4.31)
$$

Clearly, the 1st-order correction $E_n^{(1)}$ is *not* small compared with the 0th-order energy $E_n^{(0)} = n\hbar\omega$ when the quantum number n is too large. Then higher-order terms must be calculated, or a totally different method used.

6-9 Find the 2nd-order energy correction to the ground-state energy, that is: calculate $E_0^{(2)}$ for this example.

Finally a remark about the normalization of $|n\rangle$ in accordance with (6.4.5), that is $\langle n^{(0)}|n\rangle = 1$. When calculating expectation values in the nth state

$$
\langle A\rangle_n = \frac{\langle n|A|n\rangle}{\langle n|n\rangle}\,, \qquad (6.4.32)
$$

we need the factor $Z_n = \langle n|n\rangle^{-1}$ for correct normalization. Beginning with

$$
\begin{aligned}
\frac{1}{Z_n} &= \langle n|n\rangle \\
&= \Big(\langle n^{(0)}| + \langle n^{(1)}| + \langle n^{(2)}| + \cdots\Big)\Big(|n^{(0)}\rangle + |n^{(1)}\rangle + |n^{(2)}\rangle + \cdots\Big) \\
&= \underbrace{\langle n^{(0)}|n^{(0)}\rangle}_{=1\ \text{(0th-order)}} + \underbrace{\Big[\langle n^{(0)}|n^{(1)}\rangle + \langle n^{(1)}|n^{(0)}\rangle\Big]}_{=0\ \text{(1st-order)}} \\
&\quad + \underbrace{\Big[\langle n^{(0)}|n^{(2)}\rangle + \langle n^{(1)}|n^{(1)}\rangle + \langle n^{(2)}|n^{(0)}\rangle\Big]}_{=\langle n^{(1)}|n^{(1)}\rangle\ \text{(2nd-order)}} + \cdots \\
&= 1 + \langle n^{(1)}|n^{(1)}\rangle + O(H_1^3)\,, \qquad (6.4.33)
\end{aligned}
$$

we have

$$Z_n = 1 - \langle n^{(1)} | n^{(1)} \rangle + \cdots$$
$$= 1 - \sum_{m(\neq n)} \frac{\left| \langle m^{(0)} | H_1 | n^{(0)} \rangle \right|^2}{\left(E_m^{(0)} - E_n^{(0)} \right)^2} + \cdots \tag{6.4.34}$$

to second order in the perturbation, where we note in particular that there is no 1st-order term.

6-10 Regard H_0 as specified by $|n^{(0)}\rangle$ and $E_n^{(0)}$, and show that

$$Z_n = \frac{\partial E_n}{\partial E_n^{(0)}}$$

to second order (which is actually true in general).

6.5 Brillouin–Wigner perturbation theory

Rayleigh–Schrödinger perturbation theory has a tendency to become quite tedious when one wants to go beyond second order in the perturbation. It is easy to use in the first and second orders, but beyond that alternative methods may be preferable. One such method is the one by Léon Brillouin and Eugene P. Wigner.

It proceeds from the eigenvalue equation (6.4.4),

$$H|n\rangle = (H_0 + H_1)|n\rangle = |n\rangle E_n \,, \tag{6.5.1}$$

which we now write in the form

$$(E_n - H_0)|n\rangle = H_1|n\rangle \,. \tag{6.5.2}$$

Sticking to the convenient normalization of (6.4.5), $\langle n^{(0)} | n \rangle = 1$, we also have

$$|n\rangle = \sum_m |m^{(0)}\rangle\langle m^{(0)}|n\rangle$$
$$= |n^{(0)}\rangle + \sum_{m(\neq n)} |m^{(0)}\rangle\langle m^{(0)}|n\rangle \tag{6.5.3}$$

or

$$|n\rangle = |n^{(0)}\rangle + Q_n|n\rangle \tag{6.5.4}$$

with

$$Q_n = 1 - |n^{(0)}\rangle\langle n^{(0)}| = \sum_{m(\neq n)} |m^{(0)}\rangle\langle m^{(0)}| . \qquad (6.5.5)$$

This operator Q_n projects on the subspace orthogonal to $|n^{(0)}\rangle$. Since this is an eigenstate of H_0, Q_n commutes with H_0,

$$H_0 Q_n = Q_n H_0 . \qquad (6.5.6)$$

We use this property in

$$\begin{aligned} Q_n H_1 |n\rangle &= Q_n (E_n - H_0)|n\rangle \\ &= (E_n - H_0) Q_n |n\rangle \end{aligned} \qquad (6.5.7)$$

and so arrive at

$$Q_n|n\rangle = (E_n - H_0)^{-1} Q_n H_1 |n\rangle . \qquad (6.5.8)$$

Together then

$$|n\rangle = |n^{(0)}\rangle + (E_n - H_0)^{-1} Q_n H_1 |n\rangle , \qquad (6.5.9)$$

which is *not* an explicit equation for $|n\rangle$. Rather it determines $|n\rangle$ implicitly because $|n\rangle$ appears on both sides of the equation. But note that the second term on the right, $\propto H_1|n\rangle$, is smaller than $|n\rangle$ on the left by one power of H_1, so that we can generate a hierarchy of approximations by iteration:

$$\begin{aligned} |n\rangle &= |n^{(0)}\rangle + (E_n - H_0)^{-1} Q_n H_1 \Big(|n^{(0)}\rangle + (E_n - H_0)^{-1} Q_n H_1 |n\rangle\Big) \\ &= |n^{(0)}\rangle + (E_n - H_0)^{-1} Q_n H_1 |n^{(0)}\rangle \\ &\qquad + \Big[(E_n - H_0)^{-1} Q_n H_1\Big]^2 |n\rangle \\ &= |n^{(0)}\rangle + (E_n - H_0)^{-1} Q_n H_1 |n^{(0)}\rangle \\ &\qquad + \Big[(E_n - H_0)^{-1} Q_n H_1\Big]^2 |n^{(0)}\rangle \\ &\qquad + \Big[(E_n - H_0)^{-1} Q_n H_1\Big]^3 |n^{(0)}\rangle + \cdots , \end{aligned} \qquad (6.5.10)$$

which eventually leads us to the formal series

$$|n\rangle = \sum_{k=0}^{\infty} \Big[(E_n - H_0)^{-1} Q_n H_1\Big]^k |n^{(0)}\rangle . \qquad (6.5.11)$$

This is a geometric series with the sum

$$|n\rangle = \left[1 - (E_n - H_0)^{-1} Q_n H_1\right]^{-1} |n^{(0)}\rangle . \tag{6.5.12}$$

6-11 Show that this can also be presented as

$$|n\rangle = \frac{E_n - E_n^{(0)}}{E_n - H_0 - Q_n H_1} |n^{(0)}\rangle$$

where $\frac{1}{\text{operator}} = (\text{operator})^{-1}$ for brevity.

These results are charmingly compact, but they are not as useful as they appear because they involve the unknown exact energy E_n on the right. In addition, it can be forbiddingly difficult to calculate the inverse of a complicated operator.

The energy E_n can be found in a systematic approximate way by truncating the series. We have

$$\begin{aligned} E_n - E_n^{(0)} &= \langle n^{(0)} | (E_n - H_0) | n \rangle \\ &= \langle n^{(0)} | H_1 | n \rangle \end{aligned} \tag{6.5.13}$$

where we remember that $\langle n^{(0)} | n \rangle = 1$, and upon inserting the series (6.5.10) into

$$E_n = E_n^{(0)} + \langle n^{(0)} | H_1 | n \rangle , \tag{6.5.14}$$

we have

$$\begin{aligned} E_n = E_n^{(0)} &+ \langle n^{(0)} | H_1 | n^{(0)} \rangle \\ &+ \langle n^{(0)} | H_1 (E_n - H_0)^{-1} Q_n H_1 | n^{(0)} \rangle \\ &+ \langle n^{(0)} | H_1 \left[(E_n - H_0)^{-1} Q_n H_1\right]^2 | n^{(0)} \rangle \\ &+ \cdots . \end{aligned} \tag{6.5.15}$$

Truncation after the first order gives

$$E_n = E_n^{(0)} + \langle n^{(0)} | H_1 | n^{(0)} \rangle , \tag{6.5.16}$$

the familiar 1st-order result of both the Hellmann–Feynman theorem and

Rayleigh–Schrödinger perturbation theory. In second order we get

$$\begin{aligned} E_n = E_n^{(0)} &+ \langle n^{(0)} | H_1 | n^{(0)} \rangle \\ &+ \langle n^{(0)} | H_1 (E_n - H_0)^{-1} Q_n H_1 | n^{(0)} \rangle . \end{aligned} \tag{6.5.17}$$

This will look more familiar after we make use of

$$\begin{aligned} (E_n - H_0)^{-1} Q_n &= (E_n - H_0)^{-1} \sum_{m(\neq n)} |m^{(0)}\rangle\langle m^{(0)}| \\ &= \sum_{m(\neq n)} \frac{|m^{(0)}\rangle\langle m^{(0)}|}{E_n - E_m^{(0)}} , \end{aligned} \tag{6.5.18}$$

when we get

$$E_n = E_n^{(0)} + \langle n^{(0)} | H_1 | n^{(0)} \rangle - \sum_{m(\neq n)} \frac{\left| \langle m^{(0)} | H_1 | m^{(0)} \rangle \right|^2}{E_m^{(0)} - E_n} . \tag{6.5.19}$$

This looks much like the 2nd-order expression in Rayleigh–Schrödinger perturbation theory, but there is a very crucial difference. We now have E_n in the denominator, not $E_n^{(0)}$ as we did in (6.4.21).

One could argue that, since we have truncated at the third order, the answer is only consistent to second order and that, therefore, we do not lose anything by the replacement $E_n \to E_n^{(0)}$ in the final expression. There is something to an argument of this kind, but nevertheless a better approximation is often obtained by leaving E_n in the 2nd-order expression and regarding it as an implicit equation for E_n. In this manner, one gets, of course, the second order correctly but also consistent pieces of the of the higher-order terms, as they are implied at the 2nd-order stage.

A simple example is probably more convincing evidence than a sophisticated general argument. Consider, therefore, the particularly simple situation in which H_1 has only two nonvanishing matrix elements in the $|n^{(0)}\rangle$ basis of the unperturbed eigenkets,

$$\langle 0^{(0)} | H_1 | 1^{(0)} \rangle = \epsilon = \epsilon^* = \langle 1^{(0)} | H_1 | 0^{(0)} \rangle , \tag{6.5.20}$$

and $\langle m^{(0)} | H_1 | n^{(0)} \rangle = 0$ for all other choices of m and n. In the matrix representation in which H_0 is diagonal,

$$H_0 \mathrel{\widehat{=}} \begin{pmatrix} E_0^{(0)} & 0 & 0 & 0 & \cdots \\ 0 & E_1^{(0)} & 0 & 0 & \cdots \\ 0 & 0 & E_2^{(0)} & 0 & \cdots \\ 0 & 0 & 0 & E_3^{(0)} & \cdots \\ \vdots & \vdots & \vdots & \vdots & \ddots \end{pmatrix}, \tag{6.5.21}$$

we then have

$$H_1 \mathrel{\widehat{=}} \begin{pmatrix} 0 & \epsilon & 0 & \cdots \\ \epsilon & 0 & 0 & \cdots \\ 0 & 0 & 0 & \cdots \\ \vdots & \vdots & \vdots & \ddots \end{pmatrix} \tag{6.5.22}$$

and it is easy to determine the eigenvalues of

$$H = H_0 + H_1 \mathrel{\widehat{=}} \begin{pmatrix} E_0^{(0)} & \epsilon & 0 & 0 & \cdots \\ \epsilon & E_1^{(0)} & 0 & 0 & \cdots \\ 0 & 0 & E_2^{(0)} & 0 & \cdots \\ 0 & 0 & 0 & E_3^{(0)} & \cdots \\ \vdots & \vdots & \vdots & \vdots & \ddots \end{pmatrix} \tag{6.5.23}$$

namely $E_n = E_n^{(0)}$ for $n = 2, 3, 4, \ldots$, and E_0, E_1, are the solutions of

$$\det\left\{\begin{pmatrix} E_0^{(0)} - E & \epsilon \\ \epsilon & E_1^{(0)} - E \end{pmatrix}\right\} = 0\,, \tag{6.5.24}$$

that is

$$\left(E_0^{(0)} - E\right)\left(E_1^{(0)} - E\right) = \epsilon^2\,, \tag{6.5.25}$$

giving

$$\left.\begin{matrix} E_0 \\ E_1 \end{matrix}\right\} = \frac{1}{2}\left(E_0^{(0)} + E_1^{(0)}\right) \mp \sqrt{\frac{1}{4}\left(E_0^{(0)} - E_1^{(0)}\right)^2 + \epsilon^2}\,. \tag{6.5.26}$$

Rayleigh–Schrödinger perturbation theory would result in the 2nd-order approximation

$$\left.\begin{matrix} E_0 \\ E_1 \end{matrix}\right\} \cong \frac{1}{2}\left(E_0^{(0)} + E_1^{(0)}\right) \mp \frac{1}{2}\left(E_1^{(0)} - E_0^{(0)}\right)\left[1 + \frac{2\epsilon^2}{\left(E_0^{(0)} - E_1^{(0)}\right)^2}\right] \tag{6.5.27}$$

or

$$E_0 \cong E_0^{(0)} - \frac{\epsilon^2}{\left(E_1^{(0)} - E_0^{(0)}\right)},$$

$$E_1 \cong E_1^{(0)} + \frac{\epsilon^2}{\left(E_1^{(0)} - E_0^{(0)}\right)}. \tag{6.5.28}$$

By contrast, the Brillouin–Wigner formalism gives, to second order,

$$E_0 = E_0^{(0)} - \frac{\epsilon^2}{\left(E_1^{(0)} - E_0\right)} \quad \text{or} \quad \left(E_0 - E_0^{(0)}\right)\left(E_0 - E_1^{(0)}\right) = \epsilon^2 \tag{6.5.29}$$

and

$$E_1 = E_1^{(0)} - \frac{\epsilon^2}{\left(E_0^{(0)} - E_1\right)} \quad \text{or} \quad \left(E_1 - E_1^{(0)}\right)\left(E_1 - E_0^{(0)}\right) = \epsilon^2, \tag{6.5.30}$$

which in both cases is the quadratic equation (6.5.25) that gave us the exact eigenenergies in (6.5.26). In this example, then, the 2nd-order Brillouin–Wigner result is much better than the corresponding Rayleigh–Schrödinger approximation. This is particularly important when ϵ is not tiny on the scale set by $E_1^{(0)} - E_0^{(0)}$.

6-12 Consider

$$H = \underbrace{\hbar\omega A^\dagger A}_{=H_0} + \underbrace{\hbar\Omega\left({A^\dagger}^2 + A^2\right)}_{=H_1}.$$

Find the ground-state energy to second order in Ω, both by the Rayleigh–Schrödinger and by the Brillouin–Wigner method. Compare with the exact ground-state energy. Hint: Express H in terms of X and P for getting the exact energies.

6.6 Perturbation theory for degenerate states

The Rayleigh–Schrödinger perturbation formalism and the Brillouin–Wigner perturbation formalism, as presented above, assume that the unperturbed energies, that is: the eigenvalues of H_0, are not degenerate,

$$E_0^{(0)} < E_1^{(0)} < E_2^{(0)} < \cdots. \tag{6.6.1}$$

We know, however, from the examples of the two-dimensional harmonic oscillator and the three-dimensional Coulomb problem that it is quite common to have degenerate excited states. For spherically symmetric situations, there is in fact always the energy degeneracy in the L_3 quantum number m, because m does not appear in the radial Schrödinger equation (4.4.24) that determines the eigenvalues of the Hamilton operator for given angular momentum quantum numbers l and m; recall also the lesson of Exercise 4-10 on page 120.

The complications resulting from a degeneracy can be appreciated already at the level of the Hellmann–Feynman theorem. We return to (6.1.3) and (6.1.4) and make the additional quantum numbers explicit,

$$H_\lambda|E_\lambda,\gamma\rangle = |E_\lambda,\gamma\rangle E_\lambda\,,$$
$$\frac{\partial H_\lambda}{\partial\lambda}|E_\lambda,\gamma\rangle + H_\lambda\frac{\partial|E_\lambda,\gamma\rangle}{\partial\lambda} = \frac{\partial|E_\lambda,\gamma\rangle}{\partial\lambda}E_\lambda + |E_\lambda,\gamma\rangle\frac{\partial E_\lambda}{\partial\lambda}\,, \qquad (6.6.2)$$

where the symbol γ stands for the other quantum numbers (such as lm) that we indicated by "..." earlier. Now multiply the second equation by bra $\langle E_\lambda,\gamma'|$ with the same energy E_λ but perhaps different γ. This gives

$$\begin{aligned}\langle E_\lambda,\gamma'|\frac{\partial H_\lambda}{\partial\lambda}|E_\lambda,\gamma\rangle &= \frac{\partial E_\lambda}{\partial\lambda}\langle E_\lambda,\gamma'|E_\lambda,\gamma\rangle \\ &= \frac{\partial E_\lambda}{\partial\lambda}\delta(\gamma,\gamma')\,, \end{aligned}\qquad (6.6.3)$$

where we have the general version of Leopold Kronecker's delta symbol,

$$\delta(\gamma,\gamma') = \begin{cases}1 & \text{if}\quad \gamma=\gamma'\,,\\ 0 & \text{if}\quad \gamma\neq\gamma'\,.\end{cases} \qquad (6.6.4)$$

In (6.6.3), it states the orthogonality of the kets $|E,\gamma\rangle$ for different γs.

The appearance of $\delta(\gamma,\gamma')$ on the right implies that the left-hand side must also vanish for $\gamma\neq\gamma'$,

$$\langle E_\lambda,\gamma'|\frac{\partial H_\lambda}{\partial\lambda}|E_\lambda,\gamma\rangle = \begin{cases}\langle E_\lambda,\gamma|\dfrac{\partial H_\lambda}{\partial\lambda}|E_\lambda,\gamma\rangle & \text{for}\quad \gamma=\gamma'\,,\\ 0 & \text{for}\quad \gamma\neq\gamma'\,,\end{cases} \qquad (6.6.5)$$

which is really a statement about the fitting choice for the unperturbed, degenerate eigenstates of H_0 in $H = H_0+H_1$. For the purpose of perturbation theory we need to have

$$\langle n^{(0)},\gamma|H_1|n^{(0)},\gamma'\rangle = 0 \quad \text{if}\quad \gamma\neq\gamma'\,. \qquad (6.6.6)$$

In other words: The unperturbed states are to be chosen such that the perturbation H_1 has a diagonal matrix for them.

All of this is more easily understood after considering an example. We take one of particular physical interest, namely the change in the energy of hydrogen that results from an external electric field.

6.7 Linear Stark effect

We can safely assume that the electric field $\vec{E}$ does not vary significantly over the volume of the atom. In effect, we thus deal with a homogeneous electric field $\vec{E}$ that acts in addition to the Coulomb field of the nuclear charge Ze. The additional force on the electron (charge $-e<0$) is

$$\vec{F} = -e\vec{E}\,, \tag{6.7.1}$$

and the Hamilton operator of (5.1.1) acquires an extra term,

$$H = \underbrace{\frac{1}{2M}\vec{P}^2 - \frac{Ze^2}{|\vec{R}|}}_{=H_0} \underbrace{-\vec{F}\cdot\vec{R}}_{=H_1}\,, \tag{6.7.2}$$

where we continue to neglect the small correction associated with the large but finite mass of the nucleus.

We recall that the eigenvalues of H_0 are the Bohr energies of (5.1.16)

$$E_n^{(0)} = -\frac{Z^2e^2/a_0}{2n^2}\,, \qquad n=1,2,3,\ldots \tag{6.7.3}$$

which are n^2-fold degenerate inasmuch as there are n^2 states

$$\bigl|n,l,m\bigr\rangle \equiv \bigl|E_n^{(0)},\underset{\substack{\downarrow \\ lm}}{\gamma}\bigr\rangle \tag{6.7.4}$$

for the given n, with $l=0,1,2,\ldots$ and $m=0,\pm1,\pm2,\ldots,\pm l$. Only the $n=1$ ground state $\bigl|1,0,0\bigr\rangle$ is not degenerate, and the electric field has no

1st-order effect on it:

$$\begin{aligned}\langle 1,0,0|H_1|1,0,0\rangle &= \int (\mathrm{d}\vec{r})\left(-\vec{F}\cdot\vec{r}\right)\left|\langle\vec{r}|1,0,0\rangle\right|^2\\ &= \int (\mathrm{d}\vec{r})\left(-\vec{F}\cdot\vec{r}\right)\left|R_{10}(r)Y_{00}(\theta,\phi)\right|^2\\ &= \int (\mathrm{d}\vec{r})\left(-\vec{F}\cdot\vec{r}\right)\frac{1}{\pi}\left(\frac{Z}{a_0}\right)^3 \mathrm{e}^{-2Zr/a_0} = 0\,. \end{aligned} \tag{6.7.5}$$

Here we use $\langle\vec{r}|1,0,0\rangle = R_{10}(r)Y_{lm}(\theta,\phi)$ with

$$\begin{aligned} R_{10}(r) &= 2\left(\frac{Z}{a_0}\right)^{3/2} \mathrm{L}_0^{(1)}\left(\frac{2Zr}{a_0}\right)\mathrm{e}^{-Zr/a_0}\\ &= 2\left(\frac{Z}{a_0}\right)^{3/2}\mathrm{e}^{-Zr/a_0} \end{aligned} \tag{6.7.6}$$

from (5.2.28) and $Y_{00}(\theta,\phi) = (4\pi)^{-1/2}$ from (5.2.39).

For the set of first excited states, those with $n = 2$, we have $(l,m) = (0,0),(1,0),(1,1),(1,-1)$, the $2^2 = 4$ orthogonal states with definite L_3 and $\vec{L}^2$ values, so that there are altogether $4^2 = 16$ matrix elements of the form

$$\langle 2,l,m|H_1|2,l',m'\rangle = \int (\mathrm{d}\vec{r})\langle 2,l,m|\vec{r}\rangle\left(-\vec{F}\cdot\vec{r}\right)\langle\vec{r}|2,l',m'\rangle \tag{6.7.7}$$

that we need to evaluate. For this purpose, we will, of course, choose the coordinate system conveniently, namely such that the third cartesian axis that is singled out by the spherical coordinates coincides with the direction of the electric field, of the force. Then

$$\vec{F}\cdot\vec{r} = Fr\cos\theta \tag{6.7.8}$$

and we have

$$\begin{aligned}
&\langle 2,l,m|H_1|2,l',m'\rangle \\
&\quad = \int_0^\infty \mathrm{d}r\, r^2 \int_0^{2\pi} \mathrm{d}\phi \int_0^\pi \mathrm{d}\theta \sin\theta \\
&\qquad \times [R_{2l}(r)Y_{lm}(\theta,\phi)]^*(-Fr\cos\theta)[R_{2l'}(r)Y_{l'm'}(\theta,\phi)] \\
&\quad = -F \int_0^\infty \mathrm{d}r\, r^3\, R_{2l}(r)^* R_{2l'}(r) \\
&\qquad \times \underbrace{\int_0^{2\pi} \mathrm{d}\phi \int_0^\pi \mathrm{d}\theta\, \sin\theta\cos\theta\, Y_{lm}(\theta,\phi)^* Y_{l'm'}(\theta,\phi)}_{\text{angular part}}\,. \qquad (6.7.9)
\end{aligned}$$

We take a close look at the angular part first.

Recall that the ϕ dependence of Y_{lm} is given by a factor $\mathrm{e}^{\mathrm{i}m\phi}$, so that

$$\text{angular part} \propto \int_0^{2\pi} \mathrm{d}\phi\, \mathrm{e}^{-\mathrm{i}(m-m')\phi} = 2\pi\delta_{mm'}\,, \qquad (6.7.10)$$

that is: all terms with $m \neq m'$ vanish, a benefit of the well chosen coordinate system. So we only need to consider the angular momentum quantum numbers of the combinations marked by a $*$ in the following table:

$l\,m$ \ $l'm'$	00	10	11	1−1
0 0	*	*	0	0
1 0	*	*	0	0
1 1	0	0	*	0
1−1	0	0	0	*

(6.7.11)

We note that, see (5.2.39),

$$Y_{00} = \sqrt{\frac{1}{4\pi}}\,, \qquad Y_{10} = \sqrt{\frac{3}{4\pi}}\cos\theta\,,$$
$$Y_{1\pm1} = \mp\sqrt{\frac{3}{8\pi}}\,\mathrm{e}^{\pm\mathrm{i}\phi}\sin\theta\,, \qquad (6.7.12)$$

so that the angular part is

$$\begin{aligned}
l = l' = 1\,,\ m = m' = \pm 1 : \ &\frac{3}{4}\int_0^\pi \mathrm{d}\theta\, \sin\theta\cos\theta(\sin\theta)^2 \\
&= \frac{3}{16}(\sin\theta)^4\Big|_{\theta=0}^{\pi} = 0\,; \\
l = l' = 1\,,\ m = m' = 0 : \ &\frac{3}{2}\int_0^\pi \mathrm{d}\theta\, \sin\theta\cos\theta(\cos\theta)^2 \\
&= -\frac{3}{8}(\cos\theta)^4\Big|_{\theta=0}^{\pi} = 0\,; \\
l = l' = 0\,,\ m = m' = 0 : \ &\frac{1}{2}\int_0^\pi \mathrm{d}\theta\, \sin\theta\cos\theta \\
&= \frac{1}{4}(\sin\theta)^2\Big|_{\theta=0}^{\pi} = 0\,;
\end{aligned} \tag{6.7.13}$$

and finally,

$$\begin{aligned}
l = 0\,,\ l' = 1, m = m' = 0 : \ &\frac{\sqrt{3}}{2}\int_0^\pi \mathrm{d}\theta\, \sin\theta\cos\theta\cos\theta \\
&= -\frac{1}{2\sqrt{3}}(\cos\theta)^3\Big|_{\theta=0}^{\pi} = \frac{1}{\sqrt{3}}\,.
\end{aligned} \tag{6.7.14}$$

So, of the 16 matrix elements all but two are 0 as a result of their vanishing angular parts, and we need to evaluate the radial integral only for $l = 0$, $m = 0$; $l' = 1$, $m' = 0$ and $l = 1$, $m = 0$; $l' = 0$, $m' = 0$, which in fact is the same integration, see (6.7.9),

$$\begin{aligned}
\langle 2,1,0|H_1|2,0,0\rangle &= \langle 2,0,0|H_1|2,1,0\rangle^* \\
&= -F\int_0^\infty \mathrm{d}r\, r^3\, R_{21}(r)^* R_{20}(r)\frac{1}{\sqrt{3}}\,.
\end{aligned} \tag{6.7.15}$$

With

$$\begin{aligned}
R_{21}(r) &= \frac{1}{\sqrt{3}}\left(\frac{Z}{2a_0}\right)^{3/2} s\,\mathrm{e}^{-\frac{1}{2}s}\,, \\
R_{20}(r) &= \left(\frac{Z}{2a_0}\right)^{3/2} (s-2)\,\mathrm{e}^{-\frac{1}{2}s}\,,
\end{aligned} \tag{6.7.16}$$

where $s \equiv Zr/a_0$, we have

$$\begin{aligned}\langle 2,1,0|H_1|2,0,0\rangle &= -\frac{1}{3}F\frac{a_0}{Z}\frac{1}{8}\int_0^\infty \mathrm{d}s\, s^4(s-2)\,\mathrm{e}^{-s} \\ &= -\frac{1}{24}\frac{Fa_0}{Z}\underbrace{(5!-2\cdot 4!)}_{=(5-2)\cdot 4! = 3\cdot 24} \\ &= -\frac{3Fa_0}{Z}\,. \end{aligned} \tag{6.7.17}$$

The complete table (6.7.11) is therefore

$n = 2$: (6.7.18)

$l\,m$ \ lm'	00	10	11	1−1
0 0	0	$-3Fa_0/Z$	0	0
1 0	$-3Fa_0/Z$	0	0	0
1 1	0	0	0	0
1−1	0	0	0	0

This tells us that, in the $n = 2$ sector,

$$\begin{aligned} H_1|2,0,0\rangle &= |2,1,0\rangle(-3Fa_0/Z)\,, \\ H_1|2,1,0\rangle &= |2,0,0\rangle(-3Fa_0/Z)\,, \\ H_1|2,1,\pm 1\rangle &= 0\,. \end{aligned} \tag{6.7.19}$$

As a consequence $|2,1,1\rangle$ and $|2,1,-1\rangle$ are eigenstates of $H_1 = -\vec{F}\cdot\vec{R}$ with eigenvalue 0, whereas the normalized superposition states,

$$\frac{1}{\sqrt{2}}\big(|2,0,0\rangle \pm |2,1,0\rangle\big) \tag{6.7.20}$$

are eigenkets of H_1 with eigenvalues $\mp 3Fa_0/Z$,

$$H_1\frac{1}{\sqrt{2}}\big(|2,0,0\rangle \pm |2,1,0\rangle\big) = \frac{1}{\sqrt{2}}\big(|2,0,0\rangle \pm |2,1,0\rangle\big)(\mp 3Fa_0/Z)\,. \tag{6.7.21}$$

We summarize the *linear Stark effect*, which is what this phenomenon is called in recognition of Johannes Stark's contribution, graphically in a sketch:

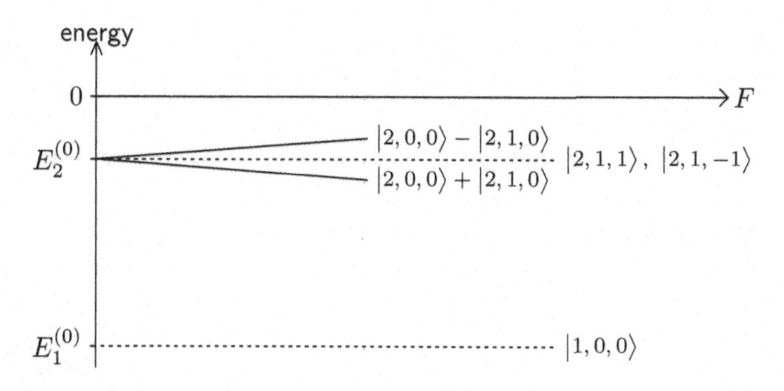

The energies of $|1,0,0\rangle$, $|2,1,1\rangle$, and $|2,1,-1\rangle$ do not change when the electric field is applied, but those of $|2,0,0\rangle \pm |2,1,0\rangle$ decrease/increase by $\mp 3Fa_0/Z$, respectively.

The energy change $3Fa_0/Z$ should be compared with the electrostatic energy of the Coulomb field at the typical distance of $r \sim a_0/Z$:

$$\frac{Fa_0/Z}{Ze^2/\frac{a_0}{Z}} = \frac{Fa_0^2}{Z^3e^2} = \frac{|\vec{E}|}{Z^3e/a_0^2} \tag{6.7.22}$$

where $|\vec{E}|$ is the strength of the electric field that is applied. This field is "weak" in the sense that 1st-order perturbation theory applies if

$$|\vec{E}| \ll \frac{Z^3e}{a_0^2} = \frac{Ze}{(a_0/Z)^2}\,. \tag{6.7.23}$$

We recall $e^2/a_0 = 27.2\,\text{eV}$ or $e/a_0 = 27.2\,\text{V}$ and $a_0 = 0.529\,\text{Å}$, and so note that, in view of

$$\frac{e}{a_0^2} = \frac{27.2}{0.529}\frac{\text{V}}{\text{Å}} = 5.14 \times 10^{11}\,\frac{\text{V}}{\text{m}}\,, \tag{6.7.24}$$

all laboratory electric fields are quite weak.

Note that we get this linear dependence on F only because $|2,0,0\rangle$ and $|2,1,0\rangle$ have the same unperturbed energy. A degeneracy of this kind, the same energy for different l values, is particular to the Coulomb potential (it also occurs in the three-dimensional harmonic oscillator, which is not so relevant, however), and we do not have a linear Stark effect in more complex atoms, helium being the simplest example, where the l-degeneracy does not exist.

In the hydrogen ground state we have a quadratic Stark effect,

$$E_0^{(2)} = -\frac{1}{2}\alpha \vec{E}^2 \tag{6.7.25}$$

with $\alpha > 0$, because the ground-state energy is *always* lowered in second order of the perturbation, as we found in (6.4.22). The energy change δE that results from a small change of the electric field, $\delta\vec{E}$, is quite generally of the form

$$\delta E = -\vec{d}\cdot\delta\vec{E}\,, \tag{6.7.26}$$

where $\vec{d}$ is the *electric dipole moment.* For the 2nd-order Stark effect, this is

$$\vec{d} = \alpha\vec{E}\,, \tag{6.7.27}$$

that is: the dipole moment itself is proportional to the applied field. The proportionality constant α is the so-called *polarizability* of the atom. There is, then, a rather simple physical picture: When applying the electric field, we induce an electric dipole moment in the atom (by redistributing the charges), and then the electric field has a "handle" on the atom, and we get an energy change as a secondary effect.

By contrast, the excited states $|2,0,0\rangle \pm |2,1,0\rangle$ of (6.7.20) have an electric dipole moment to begin with. The electric field can then act immediately and give us an energy change in first order.

6-13 Calculate the polarizability of hydrogenic atoms in the ground state by using the Rayleigh–Schrödinger perturbation theory.

6.8 WKB approximation

Consider a particle moving along the x axis, with the motion governed by a Hamilton operator of the typical form

$$H = \frac{1}{2M}P^2 + V(X) \tag{6.8.1}$$

where $V(X)$ is some reasonable potential energy function.

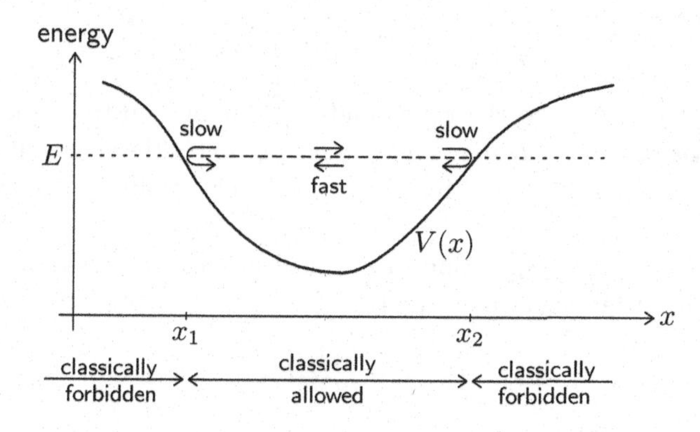

If the situation is as sketched, *classical* motion with energy E is restricted to the range $x_1 \le x \le x_2$, that is to x values between the classical turning points x_1 and x_2, where the potential energy is equal to the total energy

$$V(x_1) = E = V(x_2), \tag{6.8.2}$$

so that the kinetic energy $p^2/(2M) = E - V(x)$ vanishes. Between the turning points we have

$$E - V(x) > 0 \quad \text{for} \quad x_1 < x < x_2 \tag{6.8.3}$$

so that a positive kinetic energy results, but to the left of x_1 and to the right of x_2, we would have negative kinetic energy which makes these regions inaccessible to classical motion.

Near the turning points the kinetic energy is small, so that the velocity is small and the particle spends a relatively longer time near the turning points. A random snapshot would therefore have a good chance of showing the particle near one of the turning points, and a smaller chance of showing it in between, where the motion is faster. When we translate these considerations into the language of quantum mechanics, we expect that the probability density $|\psi(x)|^2$ is large near the classical turning points, smaller between them, and very small (exponentially decreasing) when one moves into the classically forbidden regions.

We solve the classical statement about energy conservation,

$$E = \frac{p^2}{2M} + V(x), \tag{6.8.4}$$

for the classical momentum as a function of position,

$$p = \pm\sqrt{2M\big(E - V(x)\big)} \equiv \pm p(x) \tag{6.8.5}$$

with "+" for motion to the right, and "−" for motion to the left. This *classical position-dependent momentum* $p(x)$ is real in the classically allowed region, where $E > V(x)$, and imaginary in the classically forbidden region, where $E < V(x)$.

The corresponding quantum mechanical problem asks for the solution of the Schrödinger eigenvalue equation,

$$H\big|E\big\rangle = \big|E\big\rangle E\,, \tag{6.8.6}$$

in terms of the position wave function $\psi_E(x) = \big\langle x\big|E\big\rangle$,

$$\left[-\frac{\hbar^2}{2M}\frac{\mathrm{d}^2}{\mathrm{d}x^2} + V(x)\right]\psi_E(x) = E\psi_E(x)\,. \tag{6.8.7}$$

Let us write this as

$$-\frac{\mathrm{d}^2}{\mathrm{d}x^2}\psi_E(x) = -\psi_E''(x) = \frac{1}{\hbar^2}2M\big(E - V(x)\big)\psi_E(x)\,, \tag{6.8.8}$$

where we recognize the classical momentum at position x, $p(x)$ of (6.8.5),

$$-\psi_E''(x) = \frac{1}{\hbar^2}\big[p(x)\big]^2\psi_E(x)\,. \tag{6.8.9}$$

Now note that

in the classically allowed region: $\big[p(x)\big]^2 > 0$ and $\psi_E(x)$ is oscillatory;

in the classically forbidden region: $\big[p(x)\big]^2 < 0$ and $\psi_E(x)$ is monotonically increasing or decreasing.

With this in mind, we make the *ansatz*

$$\psi_E(x) = a(x)\cos\big(\phi(x)\big) \tag{6.8.10}$$

for the wave function in the classically allowed region, expecting that we can choose the amplitude function $a(x)$ and the phase function $\phi(x)$ real and slowly varying. The oscillatory nature of $\psi_E(x)$ should be taken care of by the oscillating cosine function. With

$$\begin{aligned}\psi_E' &= a'\cos\phi - a\phi'\sin\phi\,,\\ \psi_E'' &= \left(a'' - a\phi'^2\right)\cos\phi - \left(2a'\phi' + a\phi''\right)\sin\phi\end{aligned} \tag{6.8.11}$$

we match the $\cos\phi$ and $\sin\phi$ terms in

$$\begin{aligned}-\psi_E'' &= \left(2a'\phi' + a\phi''\right)\sin\phi - \left(a'' - a\phi'^2\right)\cos\phi\\ &= \frac{1}{\hbar^2}\left[p(x)\right]^2\psi_E = \frac{1}{\hbar^2}\left[p(x)\right]^2 a\cos\phi\end{aligned} \tag{6.8.12}$$

and get

$$a'' - a\phi'^2 = -\frac{1}{\hbar^2}p^2 a \tag{6.8.13}$$

as well as

$$2a'\phi' + a\phi'' = 0\,. \tag{6.8.14}$$

The latter has $a(x)$ as an integrating factor,

$$2aa'\phi' + a^2\phi'' = \frac{\mathrm{d}}{\mathrm{d}x}\left(a^2\phi'\right) = 0\,, \tag{6.8.15}$$

so that

$$a^2\phi' = \text{const.} \tag{6.8.16}$$

follows, and (6.8.13) says

$$\phi'^2 = \left(\frac{p(x)}{\hbar}\right)^2 + \frac{a''}{a}\,. \tag{6.8.17}$$

Now, since we are relying on ideas borrowed from classical mechanics, we should concern ourselves with (highly) excited states, that is: relatively large energy eigenvalue E. We then expect that $\psi = a\cos\phi$ undergoes many oscillations of the cosine before the amplitude $a(x)$ changes considerably. If so, the term a''/a should not be overly important, and so we arrive at the

$$\text{0th approximation: ignore } \frac{a''}{a} \text{ in (6.8.17).} \tag{6.8.18}$$

This is known as the WKB approximation, where the initials stand for Gregor Wentzel, Hendrik A. Kramers, and Léon Brillouin who, in independent pieces of research, came up with essentially the same idea at the same time. Some, mostly the British, speak of the JWKB approximation, remembering thus the somewhat earlier contribution by Sir Harold Jeffreys. In fact, the history of this subject begins much earlier with contributions from George Green and Francesco Carlini. Perhaps, then, we should advocate it as the CGJWKB approximation?

We follow their guidance and see what we get upon neglecting the a''/a term. Then

$$\phi'(x) = \pm\frac{1}{\hbar}p(x) \quad \text{with} \quad p(x) = \sqrt{2M\bigl(E - V(x)\bigr)} > 0\,, \tag{6.8.19}$$

where we can just take the + solution because eventually the sign of ϕ will not matter in $\psi_E = a\cos\phi$. So

$$\phi(x) = \frac{1}{\hbar}\int^x \mathrm{d}x'\, p(x')\,, \tag{6.8.20}$$

where a particular choice of the lower integration limit fixes the value of the integration constant, which is of no concern to us presently.

From (6.8.16) we then get

$$a(x) \propto p(x)^{-\frac{1}{2}} \tag{6.8.21}$$

with a proportionality factor that is to be determined by the normalization of $\psi_E(x)$. This detail is also not so interesting for us right now.

We expect the WKB approximation to give good results whenever a''/a is small compared to $\bigl[p(x)/\hbar\bigr]^2$, which means, roughly, that

$$\left|\frac{\hbar}{p(x)}\frac{\mathrm{d}}{\mathrm{d}x}V(x)\right| \ll E - V(x)\,. \tag{6.8.22}$$

We meet there the so-called (reduced) de Broglie wavelength

$$\lambda\!\!\!^{-}(x) = \frac{\hbar}{p(x)} = \bigl[\phi'(x)\bigr]^{-1}\,, \tag{6.8.23}$$

named after Prince Louis-Victor de Broglie. Its physical significance is revealed by recalling that

$$\cos\Bigl(2\pi\frac{x}{\lambda}\Bigr) = \cos\Bigl(\frac{x}{\lambda\!\!\!^{-}}\Bigr)\,, \quad (\lambda \text{ constant}) \tag{6.8.24}$$

is an oscillation with wavelength λ. Now, in

$$\cos\bigl(\phi(x+\mathrm{d}x)\bigr) = \cos\bigl(\phi(x) + \phi'(x)\mathrm{d}x\bigr) \tag{6.8.25}$$

we see that an increment by $\mathrm{d}x$ changes the phase by

$$\phi'(x)\mathrm{d}x = \frac{\mathrm{d}x}{\hbar/p(x)} = \frac{\mathrm{d}x}{\lambda\!\!\!^{-}(x)}\,, \tag{6.8.26}$$

and this tells us that $\lambda(x)$ plays the role of a local wavelength. Of course, this is a concept that is only meaningful if there are many oscillations before $\lambda(x)$ changes significantly. In

$$\left|\lambda(x)\frac{\mathrm{d}}{\mathrm{d}x}V(x)\right| \ll E - V(x) \tag{6.8.27}$$

it states that the WKB approximation is reliable if $V(x)$ does not change much over the distance of a few de Broglie wavelengths.

Clearly, this is not true in the immediate vicinity of the classical turning points, where $E - V(x) \cong 0$. But it is expected to be all right inside the classically allowed region, at least as long as the energy E is reasonably above the ground-state energy associated with the potential $V(x)$.

Let us now return to (6.8.20). The total phase accumulated between the turning points is

$$\phi(x_2) - \phi(x_1) = \frac{1}{\hbar}\int_{x_1}^{x_2} \mathrm{d}x\, p(x)\,. \tag{6.8.28}$$

It is reasonably plausible (and can be justified by more sophisticated arguments) that this difference takes on only particular values, reflecting thereby the particular energy eigenvalues of the Hamilton operator. Let us see what we get for the case of a harmonic oscillator, for which we know the energy eigenvalues and for which we can also evaluate the integral over $p(x)$.

Thus we take

$$V(x) = \frac{1}{2}M\omega^2x^2\,, \quad E = \hbar\omega\left(n + \frac{1}{2}\right) \tag{6.8.29}$$

and have

$$\phi(x_2) - \phi(x_1) = \frac{1}{\hbar}\int_{x_1}^{x_2} \mathrm{d}x\,\sqrt{2ME - (M\omega x)^2} \tag{6.8.30}$$

with

$$\left.\begin{matrix} x_1 \\ x_2 \end{matrix}\right\} = \mp\frac{\sqrt{2ME}}{M\omega}\,. \tag{6.8.31}$$

Substitute

$$\begin{aligned} x &= \frac{\sqrt{2ME}}{M\omega}\sin\vartheta\,, \qquad -\frac{\pi}{2} \le \vartheta \le \frac{\pi}{2}\,, \\ \mathrm{d}x &= \frac{\sqrt{2ME}}{M\omega}\mathrm{d}\vartheta\,\cos\vartheta\,, \\ \sqrt{2ME - (M\omega x)^2} &= \sqrt{2ME}\cos\vartheta \end{aligned} \tag{6.8.32}$$

and get

$$\begin{aligned} \phi(x_2) - \phi(x_1) &= \frac{1}{\hbar}\frac{2E}{\omega}\underbrace{\int_{-\pi/2}^{\pi/2}\mathrm{d}\vartheta(\cos\vartheta)^2}_{=\pi/2} \\ &= \pi\frac{E}{\hbar\omega} = \pi\left(n + \frac{1}{2}\right). \end{aligned} \tag{6.8.33}$$

Taking this lesson about the harmonic oscillator as guidance, we require more generally $\phi(x_2) - \phi(x_1) = (n + \frac{1}{2})\pi$ when E is the nth energy eigenstate, that is

$$\frac{1}{\pi\hbar}\int_{x_1}^{x_2}\mathrm{d}x\,\sqrt{2ME_n - 2MV(x)} = n + \frac{1}{2}\,, \tag{6.8.34}$$

which is, indeed, the correct *WKB quantization rule.* It determines the WKB estimates for the energy eigenvalues by putting $n = 0, 1, 2, \ldots$ on the right-hand side. By construction, it gives the correct eigenvalues for the harmonic oscillator.

A further plausibility argument requests that the turning points are on equal footing, which translates into the requirement

$$\cos\bigl(\phi(x_1)\bigr) = \pm\cos\bigl(\phi(x_2)\bigr)\,. \tag{6.8.35}$$

In conjunction with the quantization rule (6.8.34), this gives

$$\begin{aligned} \cos\bigl(\phi(x_2)\bigr) &= \cos\bigl(\phi(x_1) + (n + \tfrac{1}{2})\pi\bigr) \\ &= (-1)^n\cos\bigl(\phi(x_1) + \tfrac{1}{2}\pi\bigr) \\ &= \pm\cos\bigl(\phi(x_1)\bigr) \end{aligned} \tag{6.8.36}$$

which is met if $\phi(x_1) = -\frac{\pi}{4}$. Together then

$$\phi(x) = \frac{1}{\hbar}\int_{x_1}^{x} \mathrm{d}x'\, p(x') - \frac{\pi}{4}$$
$$\text{with}\quad p(x) = \sqrt{2ME\bigl(E - V(x)\bigr)}\,. \tag{6.8.37}$$

As an illustration of the WKB quantization rule (6.8.34), we try it out for $V(x) = F|x|$ with $F > 0$, the potential of the constant restoring force of (6.3.5). The turning points are at

$$\left.\begin{matrix} x_1 \\ x_2 \end{matrix}\right\} = \mp E_n/F\,, \tag{6.8.38}$$

so that

$$\begin{aligned} n + \frac{1}{2} &= \frac{2}{\pi\hbar}\int_0^{E_n/F} \mathrm{d}x\,\sqrt{2M(E_n - Fx)} \\ &= \frac{2}{\pi\hbar}\frac{-1}{3MF}\left[2M(E_n - Fx)\right]^{3/2}\Big|_{x=0}^{E_n/F} \\ &= \frac{2}{\pi\hbar}\frac{(2ME_n)^{3/2}}{3MF}\,, \end{aligned} \tag{6.8.39}$$

with the consequence

$$E_n = \frac{1}{2}\left[\frac{3\pi}{2}\left(n + \frac{1}{2}\right)\right]^{2/3}\left(\frac{\hbar^2 F^2}{M}\right)^{1/3}. \tag{6.8.40}$$

We compare the exact values for $E_n/\bigl(\hbar^2F^2/M\bigr)^{1/3}$ with the WKB approximate values in this table:

n	exact	WKB	error
0	0.8086	0.8853	+9.5%
1	1.8558	1.8416	−0.77%
2	2.5781	2.5888	+0.41%
3	3.2446	3.2397	−0.15%
4	3.8257	3.8306	+0.13%
5	4.3817	4.3790	−0.006%

The WKB approximation is stunningly good, unreasonably good one might say, except for the ground state. But for estimating the ground-state energy, we have different methods at our disposal, such as the variational estimates of the Rayleigh–Ritz method.

There is also a WKB quantization rule for spherically symmetric three-dimensional potential energy functions $V(|\vec{r}|) = V(r)$. We return to the differential equation (4.4.24),

$$\left[-\frac{\hbar^2}{2M}\frac{\mathrm{d}^2}{\mathrm{d}x^2} + \frac{\hbar^2 l(l+1)}{2Mr^2} + V(r)\right] r\psi_{n_r l} = E_{n_r l}\, r\psi_{n_r l}\,, \qquad (6.8.41)$$

and try to read it as a one-dimensional Schrödinger eigenvalue equation with the effective potential

$$V(r) + \frac{\hbar^2 l(l+1)}{2Mr^2}\,. \qquad (6.8.42)$$

This does not really work well, however, and there is a reason for this failure: the range of r is $r > 0$, whereas that of x in the one-dimensional WKB argument is $-\infty < x < \infty$. Thus, before invoking the WKB argument, we should convert the radial equation for r into one for an auxiliary x variable that ranges over all real values. As noted by Rudolph E. Langer, the correct procedure is to write $r = \mathrm{e}^x$, then derive the differential equation in x, make it look like a one-dimensional Schrödinger eigenvalue equation, apply WKB to it, and finally return to r as the integration variable. When all of this is done, the outcome is the following *WKB quantization rule for spherically symmetric three-dimensional potentials*:

$$n_r + \frac{1}{2} = \frac{1}{\pi\hbar}\int \mathrm{d}r \sqrt{2M\left(E_{n_r l} - V_l(r)\right)} \qquad (6.8.43)$$

with

$$V_l(r) = \frac{\hbar^2\left(l+\frac{1}{2}\right)^2}{2Mr^2} + V(r)\,. \qquad (6.8.44)$$

The net effect of Langer's reasoning is the replacement

$$l(l+1) \to \left(l+\tfrac{1}{2}\right)^2 \qquad (6.8.45)$$

which does not amount to much if l is large, but improves matters dramatically for small l values.

6-14 Apply the three-dimensional WKB quantization rule to

$$V(r) = \frac{1}{2}M\omega^2 r^2$$

and to

$$V(r) = -\frac{Ze^2}{r}.$$

In both cases, you should get the exact eigenvalues.

As a final remark about WKB, now returning to the basic one-dimensional situation, we consider a very different approach, which is, however, in the same semiclassical spirit. For that, we take for granted that the eigenvalues of $H = P^2/(2M) + V(X)$ are nondegenerate and ordered naturally,

$$-\infty < E_0 < E_1 < E_2 < \cdots . \tag{6.8.46}$$

We introduce a function that counts the number of eigenvalues below a parameter energy $\mathcal{E}$,

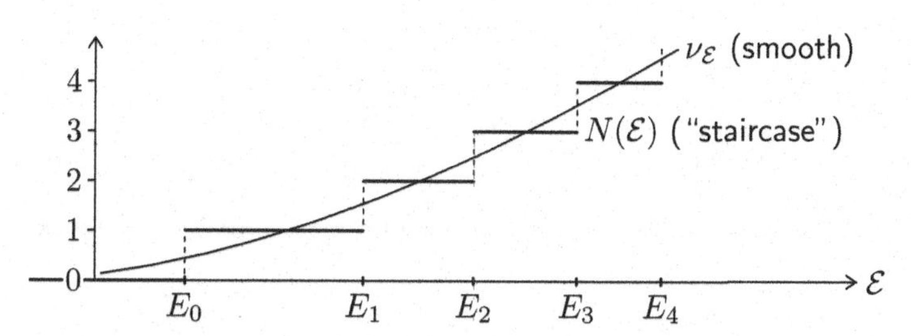

$$N(\mathcal{E}) = \{\text{number of } E_n \text{ such that } E_n < \mathcal{E}\} = \sum_n \eta(\mathcal{E} - E_n) \tag{6.8.47}$$

where

$$\eta(x) = \begin{cases} 1 & \text{for} \quad x > 0, \\ 0 & \text{for} \quad x < 0, \end{cases} \tag{6.8.48}$$

is Oliver Heaviside's unit step function, an antiderivative of Dirac's delta function $\delta(x)$. Exploiting the orthonormality and completeness of the eigenkets $|E_n\rangle$ of H,

$$\langle E_n | E_m \rangle = \delta_{nm}, \quad H = \sum_n |E_n\rangle E_n \langle E_n|, \tag{6.8.49}$$

we note that

$$N(\mathcal{E}) = \mathrm{tr}\{\eta(\mathcal{E} - H)\} . \tag{6.8.50}$$

This trace can be evaluated by a phase-space integration

$$N(\mathcal{E}) = \int \frac{\mathrm{d}x\,\mathrm{d}p}{2\pi\hbar} \frac{\langle x|\eta(\mathcal{E} - H)|p\rangle}{\langle x|p\rangle} . \tag{6.8.51}$$

We know from Exercise 1-14 on page 33 that

$$\frac{\langle x|H|p\rangle}{\langle x|p\rangle} = \frac{p^2}{2M} + V(x) \tag{6.8.52}$$

and establish easily that

$$\begin{aligned} \frac{\langle x|H^2|p\rangle}{\langle x|p\rangle} &= \left(\frac{p^2}{2M} + V(x)\right)^2 + \frac{\langle x|\left[P^2, V(X)\right]|p\rangle}{2M\langle x|p\rangle} \\ &= \left(\frac{p^2}{2M} + V(x)\right)^2 + O(\hbar) . \end{aligned} \tag{6.8.53}$$

The inference is

$$\begin{aligned} \frac{\langle x|f(H)|p\rangle}{\langle x|p\rangle} &= f\Big(\frac{p^2}{2M} + V(x)\Big) + \{\text{terms} \propto \hbar\} + \{\text{terms} \propto \hbar^2\} + \cdots \\ &= f\Big(\frac{p^2}{2M} + V(x)\Big) + \text{“quantum corrections”} . \end{aligned} \tag{6.8.54}$$

Accordingly, we have

$$N(\mathcal{E}) = \nu_{\mathcal{E}} + \text{“quantum corrections”} \tag{6.8.55}$$

with

$$\nu_{\mathcal{E}} = \int \frac{\mathrm{d}x\,\mathrm{d}p}{2\pi\hbar}\, \eta\Big(\mathcal{E} - \frac{p^2}{2M} - V(x)\Big) . \tag{6.8.56}$$

This $\nu_{\mathcal{E}}$ is a *smooth* function of $\mathcal{E}$, it does not have the discontinuous jumps that are characteristic for $N(\mathcal{E})$, and we can safely assume that it interpolates the staircase of $N(\mathcal{E})$ in (6.8.47). This suggests that we should expect $\nu_{\mathcal{E}} \equiv n + \frac{1}{2}$ for $\mathcal{E} = E_n$. We take this suggestion serious and so arrive at

$$n + \frac{1}{2} = \int \frac{\mathrm{d}x\,\mathrm{d}p}{2\pi\hbar}\, \eta\Big(E_n - \frac{p^2}{2M} - V(x)\Big) \tag{6.8.57}$$

as an implicit approximation for E_n. The step function $\eta(\ )$ limits the values of x and p to the classically allowed values that obey $p^2/(2M)+V(x) \le E_n$. In particular, the p integration covers the values

$$-\sqrt{2M(E_n - V(x))} < p < \sqrt{2M(E_n - V(x))}\,. \tag{6.8.58}$$

Therefore,

$$n + \frac{1}{2} = \frac{1}{\pi\hbar}\int \mathrm{d}x\,\sqrt{2M(E_n - V(x))}\,, \tag{6.8.59}$$

where the x integration covers all values for which $V(x) < E_n$. In the situation that is depicted on page 173, this is the range $x_1 < x < x_2$ between the classical turning points, and then (6.8.59) is exactly the WKB quantization rule of (6.8.34).

This second derivation of (6.8.34) is much more direct and, in particular, does not involve approximations to the wave functions. Further, by working out, in a systematic manner, the "quantum corrections" of (6.8.54), one can improve on the WKB quantization rule quite systematically. These observations speak in favor of the second derivation.

6-15 For the square-well potential

$$V(x) = \begin{cases} 0 & \text{for} \quad |x| > a/2\,, \\ -V_0 < 0 & \text{for} \quad |x| < a/2\,, \end{cases}$$

it is established in *Basic Matters* that the total number N_{tot} of bound states is such that

$$N_{\mathrm{tot}} - 1 < \frac{a}{\pi\hbar}\sqrt{2MV_0} < N_{\mathrm{tot}}\,.$$

Calculate $\nu_{\mathcal{E}=0}$ for this potential, and compare.

Index

Note: Page numbers preceded by the letters BM or PE refer to *Basic Matters* and *Perturbed Evolution*, respectively.